Advances in Computer Vision and Pattern Recognition

More information about this series at http://www.springer.com/series/4205

Katsushi Ikeuchi · Yasuyuki Matsushita ·
Ryusuke Sagawa · Hiroshi Kawasaki ·
Yasuhiro Mukaigawa · Ryo Furukawa ·
Daisuke Miyazaki

Active Lighting and Its Application for Computer Vision

40 Years of History of Active Lighting Techniques

 Springer

Katsushi Ikeuchi (iD)
Applied Robotics Research
Microsoft Corporation
Redmond, WA, USA

Ryusuke Sagawa (iD)
Artificial Intelligence Research Center
National Institute of Advanced Industrial
Science and Technology
Tsukuba, Ibaraki, Japan

Yasuhiro Mukaigawa (iD)
Division of Information Science
Nara Institute of Science and Technology
Nara, Japan

Daisuke Miyazaki (iD)
Department of Intelligent Systems
Hiroshima City University
Hiroshima, Japan

Yasuyuki Matsushita (iD)
Graduate School of Information Science and
Technology
Osaka University
Osaka, Japan

Hiroshi Kawasaki (iD)
Department of Advanced Information
Technology
Kyushu University
Fukuoka, Japan

Ryo Furukawa (iD)
Faculty of Information Sciences
Hiroshima City University
Hiroshima, Japan

ISSN 2191-6586   ISSN 2191-6594 (electronic)
Advances in Computer Vision and Pattern Recognition
ISBN 978-3-030-56579-4   ISBN 978-3-030-56577-0 (eBook)
https://doi.org/10.1007/978-3-030-56577-0

This Springer imprint is published by the registered company Springer Nature Switzerland AG
The registered company address is: Gewerbestrasse 11, 6330 Cham, Switzerland

Preface

Computer vision first emerged as a sub-area of artificial intelligence, of which goal is building artificial life-forms with computers. Computer vision, for building input systems of such artificial life-forms, aims to establish computational models that allow machines to perceive the external 3D world in ways similar to human visual systems. Because humans utilize their eyes as sensors for this purpose, computer-vision researchers initially developed solutions that used passive sensors similar to human eyes (i.e., television cameras) as input devices.

Although the external world surrounding us is 3D, the images projected on each retina or a mechanical imaging device is 2D. In either system, one-dimensional reduction occurs. People learn to perceive 3D by augmenting this reduction using the so-called "common sense," developed through the evolution and learning process. During the development of computer-vision science/technology, one of the key efforts has been to formulate and to integrate the common sense into computational methods.

Physics-based vision, one school of computer vision, tries to model the common sense in image formation processes. First, the image formation process from the original 3D world to a 2D projection is formulated using rigorous mathematical models based on the optics and physics. Then, the inverse process is solved by augmenting the dimensional reduction by introducing mathematical formula to represent the common sense.

One example of the common-sense formulations in this school is Ikeuchi and Horn's "smoothness constraint" proposed in their shape-from-shading algorithm. The relationship between an observed image-brightness value and a surface orientation can be formulated as the image irradiance equation at each pixel. One observation at one pixel provides only one image irradiance equation, with two unknown parameters, the two degrees of freedom (DoF) of the surface orientation. One-dimensional deficiency occurs. The smoothness constraint is formulated as an equation from the assumption that nearby pixels have similar surface orientations. By adding this as an augmented equation and setting up two equations, one image irradiance equation and one smoothness constraint equation, at each pixel, the simultaneous equations can be solved for uniquely determining the surface orientation at that pixel.

This passive method, however, has a fundamental limitation. The passive method relies on the common-sense assumptions, which usually work well but sometimes they fail. For example, the smoothness assumptions are broken at object boundaries and the passive method based on the assumptions does not work well at object boundaries. When considering computer vision not as a science but as an engineering endeavor, this characteristic is not desirable and is even problematic for recent applications which require clarification of the scope within which the methods are guaranteed to work.

Active-lighting methods augment the dimensional reduction by actively controlling lighting. Woodham's photometric stereo technique is a historical and representative example of an active-lighting method. As described, observing one image brightness value at one pixel allows us to set up one image irradiance equation with two unknown values. Different lighting-directions provide different image-irradiance equations at that pixel when observed from the same point. By actively lighting the same object from multiple directions, photometric stereo can establish multiple image-irradiance equations at each pixel and robustly determine the surface orientation of that pixel without relying on the "weak" common sense.

It is the time to refocus on active lighting methods. Since the introduction of photometric stereo techniques, other active-lighting methods have been developed such as time-of-flight (ToF) and light-striping techniques. Traditionally, however, active lighting has not received the necessary focus and legitimate evaluation in mainstream computer-vision science. This is probably because humans do not possess their own light sources actively controlled at their will.

Active lighting is an effective way to compensate for the 3D-to-2D dimensional reduction in vision systems. Moreover, active-lighting methods can guarantee when they will work, how much accuracy they will provide, and when they will fail based on the nature of their engineering designs. The demand for reliable vision sensors is rapidly increasing in many application areas. For solutions, we do not need to be tied to human-like approaches. We do not need a method like a bird, to fly in the sky; Airplanes and rockets can fly using completely different engineered solutions.

This book explores this new augmentation of the 3D-to-2D reduction using active lighting. The Part I explains the photometric and geometric aspects of computer vision, which form the basic knowledge in understanding the principles of active-lighting sensors. Furthermore, the part describes the characteristics of those sensor devices and the light sources utilized in the measurement units of active-lighting sensors. Readers with basic knowledge of computer vision may skip this part.

The Part II is the core of the book and explains representative algorithms of active-lighting techniques, including photometric stereo and structured light. This part not only describes theory of the algorithms, but also includes detail explanation of each technique, and thus, we believe that this part will greatly help the readers to re-implement the algorithm in their own systems.

In the Part III, real applications using active lighting systems, which were researched and developed by authors, are explained in detail. We believe that such showcase of real applications help the readers to better understand the potential and future direction of active lighting techniques. Let us push forward in this promising direction with this book.

Redmond, Washington, USA   Katsushi Ikeuchi
Suita, Osaka, Japan   Yasuyuki Matsushita
Tsukuba, Ibaraki, Japan   Ryusuke Sagawa
Fukuoka, Fukuoka, Japan   Hiroshi Kawasaki
Ikoma, Nara, Japan   Yasuhiro Mukaigawa
Hiroshima, Hiroshima, Japan   Ryo Furukawa
Hiroshima, Hiroshima, Japan   Daisuke Miyazaki

Contents

Part I
Basics and Theory

Chapter 1
Photometry

Active-lighting sensors project light onto target objects or the scenes to obtain photometric measurements of the reflected light, including travel time and return angle. Therefore, understanding these sensor principles requires knowledge of the photometric characteristics of light and the mechanisms of light reflection. This chapter prepares the reader for detailed principle descriptions in later chapters by explaining various photometric characteristics of light and the definitions of brightness. Furthermore, a method for determining the relationship between the original brightness of light and the measured brightness is described.

1.1 Light and Its Physical Characteristics

One of the most important properties of "light" is the duality of waves (i.e., radio wave) and particles (i.e., photons). Active-lighting sensors leverage these dual properties, such as performing measurement using wave properties or measuring the brightness of light using particle properties.

Light has a wavelength. Humans only perceive light of a very limited wavelength, as visible light, within a wide range of the light family. This section describes this visible light and its surrounding light such as infrared light and ultraviolet light, often used in the active-lighting sensors. In addition, the mechanism of interference of light waves and polarized light, which is another characteristic of the wave, will be described.

As for the particle property, attenuation and scattering will be described.

© Springer Nature Switzerland AG 2020
K. Ikeuchi et al., *Active Lighting and Its Application for Computer Vision*,
Advances in Computer Vision and Pattern Recognition,
https://doi.org/10.1007/978-3-030-56577-0_1

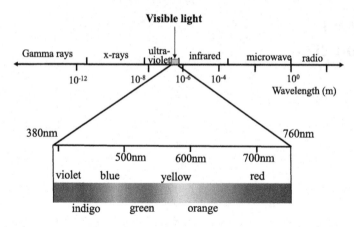

Fig. 1.1 Visible light

1.1.1 Wave and Particle

As mentioned, light possesses the properties of waves. *Visible light* comprises an electromagnetic spectrum having wavelengths between 380 and 760 nm (see Fig. 1.1). Our human visual system can only sense this class of light in a color spectrum from violet through red. This bandwidth is immediately surrounded by the shorter wavelengths of ultraviolet light on the high side and the longer wavelengths of infrared light on the low side. Some nonhuman creatures (e.g.., dogs) can naturally perceive longer infrared light waves, whereas some birds can perceive ultraviolet. Classic red-green-blue (RGB) cameras also capture near-infrared (NIR) light.

Simultaneously, light also has the properties of particles. The photoelectric effect is direct evidence of the particle nature of light. When photons interact with specific substances, photoelectrons are emitted from the energy exchange. The photoelectric effect was modeled by Einstein within his photon theory, which won him the Nobel Prize. When photoelectrons are emitted from a substance because of photon interactions, it is referred to as an external photoelectric effect. Certain imaging devices, such as photoelectric tubes and photomultipliers, utilize this effect to measure input light. When photoelectrons are released inside of a substance, it is referred to as an internal photoelectric effect. Imaging devices, such as charge-coupled-device (CCD) sensors, measure input light by detecting changes in the electronic charges within these substances.

1.1.2 Visible and Non-visible Light Family

Although humans are limited in their capacity to perceive electromagnetic waves, active sensors have access to a much wider spectrum. Those systems not only use

NIR, but they have also recently applied medium-infrared (MIR) and far-infrared (FIR) to better penetrate fog and haze. Ultraviolet light is also frequently used for fluorescence. Therefore, we cover both in this book.

200–380 nm (Ultraviolet)

Ultraviolet has its wavelength between 200 and 400 nm. Because wavelengths shorter than 320 nm (ultraviolet types B and C) are harmful to humans, wavelengths between 320 and 380 nm are commonly used as light sources. One important use of ultraviolet light is the excitation of fluorescent materials. Although the light source is itself invisible to humans, objects that include the fluorescent materials emit visible light as part of its excitation. Thus, such effects are used for forgery prevention, inspection, and advertisement purposes.

380–760 nm (RGB)

Wavelengths between 380 and 760 nm are visible as RGB colors. The human eye has three types of retinal rod cells, each having their own wavelength sensitivity corresponding to red, green, and blue. These cells enable us to perceive colors, as shown in Fig. 1.1. All colors perceivable by humans are created by the combination of the signals transmitted by these three types of rod cells to the brain via the optic nerve. Therefore, mimicking these color profiles in our technology is crucial, and thus this is one of the important purposes for developing light sources.

760–1400 nm (NIR)

NIR is practically important for 3D scanning, remote controlling, or communications. First, it is invisible to humans and does not interfere with our human visual perception. Second, although silicon (Si)-based photodetectors are commonly used for RGB, they can also be used for NIR. This is advantageous, because black objects in RGB are not truly black in NIR and can be simultaneously detected. One drawback is that NIR is almost completely absorbed by water and cannot be used for wet environments, including underwater scenes.

1400–15000 nm (MIR)

The wavelength between 1400 and 15000 nm ($= 15$ μm) includes SWIR, MWIR, and LWIR. One important feature of this wavelength is that it can go through fog and haze without severe absorption. These have recently drawn wide applications in autonomous driving technologies via light-detection and ranging (LiDAR). Si sensors cannot be used for these wavelengths. Therefore, indium (In), gallium (Ga), and arsenide (InGaAs) sensors have been effective and are now intensively applied for SWIR. Indium antimonide (InSb) has also been considered for LWIR. The clear problem with these materials is their high cost; there is no efficient or cheap way to produce large wafers with these rare-metal materials.

15–1000 μm (FIR)

Wavelength between 15 and 1000 μm is called far-infrared (FIR). Blackbody radiation theory applies here. Blackbody objects emit large quanta of FIR at low temperatures. For example, peak wavelengths emitted at 36 °C are 25 μm long. FIR sensors can, therefore, be used for thermal imaging devices that rely on temperature attributes.

1.1.3 Reflection, Refraction, and Interference

Light from a point source travels in a straight line until it hits the surface of an opaque obstacle. Figure 1.2 shows the local geometry at the point where the light hits the surface. The incoming direction is referred to as the *incident direction*. The direction perpendicular to the surface orientation of the obstacle is referred to as the *surface normal*. The angle between the incident direction and the surface normal is referred to as the *incident angle*. The light is either immediately reflected or absorbed by the substance of the obstacle. After some interactions with the internal substance, the absorbed component also re-appears outside the object. These components are observed by an observer. The direction to the observer is referred to as the *emitting direction* and the angle between the surface normal and the emitting direction is referred to as the *emitting angle*.

An example of a substance which causes immediate reflection is a mirror surface. In this case, all the incoming light is reflected and proceeds in a direction different from its incoming direction. This effect is referred to as *mirror reflection* or *specular reflection*. Let us suppose that the incoming direction, the emitting direction, and the local surface normal be $\mathbf{i}$, $\mathbf{e}$, and $\mathbf{n}$, respectively. In a perfect mirror reflection, $\mathbf{i} \cdot \mathbf{n} = \mathbf{e} \cdot \mathbf{n}$, namely the incident angle, θ_i, is equal to the emitting angle, θ_e, such that

$$\theta_i = \theta_e. \tag{1.1}$$

The three vectors, $\mathbf{i}$, $\mathbf{n}$, and $\mathbf{e}$ are co-planar,

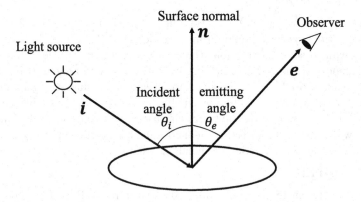

Fig. 1.2 Local geometry

$$(\mathbf{i} \times \mathbf{n}) \cdot \mathbf{e} = 0. \tag{1.2}$$

In the absorption case, the incoming light penetrates the surface of the obstacle, interacts with the internal pigments of the obstacle's substance, and eventually re-emits from the surface of the obstacle. This effect is referred to as the *diffuse reflection*. In the typical case, owing to the random nature of the interaction process, the returned energy is distributed evenly in all directions. Thus, when the surface is observed from an oblique direction, the apparent area of the emitting surface decreases. Therefore, the emitting area becomes

$$I_e = \mu I_i \cos \theta_e. \tag{1.3}$$

Here, μ, I_i and I_e are the coefficient, the incident, and emitting brightness, respectively. This is called Lambert's cosine law [1]. On the other hand, from a different observation point, the visible area per unit viewing angle increases. This effect offsets and the Lambertian surface looks the same in all directions. Details will be explained in the later chapter.

Generally, in a vacuum, the straight travel of light is guaranteed. Thus, when an object absorbs light, a distinct shadow is created behind it. Shadows in a vacuum have clear contours. Even outside a vacuum, active sensors utilize the geometry of shadows, obstacles, and the light source to determine the locations of the objects. In the air, shadow contours blur due to atmospheric fluctuation. It is possible to estimate the degree of fluctuation by analyzing the blur.

Light can also (non-gravitationally) bend around an obstacle and arrive at points that cannot be reached via a straight line. This phenomenon is called diffraction. The longer the wavelength with respect to the size of the obstacle, the larger the angle of the bend (i.e., encircling).

Light penetrates transparent substances. The speed of light in such a substance is slower than that in a vacuum or in air. Thus, when light meets the surface of a transparent object from the air, the velocity difference causes a change in the light's direction of travel. This phenomenon is referred to as *refraction*. Snell's law models this phenomenon. Referring to the volume of a substance in a transparent obstacle as a medium, if we let the propagation velocity of light in medium A to v_A, and that in medium B be v_B, we also let the incident angle from medium A to medium B be θ_A and the emitting angle from medium A to medium B be θ_B (see Fig. 1.3). We get the equation,

$$\frac{\sin \theta_A}{\sin \theta_B} = \frac{v_A}{v_B}. \tag{1.4}$$

The ratio between the velocities is defined as the relative refraction index between medium A and medium B, the equation is expressed as

$$n_{AB} = \frac{v_A}{v_B}. \tag{1.5}$$

Fig. 1.3 Refraction

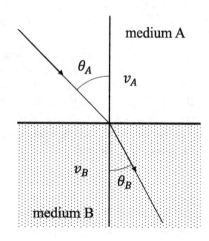

The refractive index of a medium also varies depending on the wavelength of light. The shorter the wavelength, the higher the refractive index. For example, light representing the color purple is easier to bend than red. The reason a glass prism creates a spectrum of color bands is that the various wavelengths comprising white light pass through a glass medium at different angles as it traverses the medium and end up different positions depending on the wavelengths. This is referred to as spectral phenomenon, wherein the colors are apparently separated. Rainbows are observed when sunlight refracts similarly through water droplets in the air, causing a spectral effect. A lens material can suffer from this effect. A convex lens, for example, uses refraction to collect incoming light and focus it onto a smaller space. However, a simple one-layer lens will exhibit a spectral effect and generate color distributions along the edges. To avoid these distortions, high-grade lenses achieve subtle light-path adjustments by passing light through multiple layers of glass with different refractive indices.

Interference strips are also observed due to the waves property. When multiple waves overlap, an additive wave is created. When the peak of one wave coincides with the peak of another, the absolute value of the amplitude increases (it gets brighter). When the peak of a wave coincides with the valley of another, the absolute value of the amplitude decreases (it gets dimmer). White light, such as Sunlight, contains all wavelengths of visible light, which, via the averaging-out of multiple wavelengths, minimizes the interference strips. On the other hand, monochromatic light sources (e.g., light amplification by stimulated emission of radiation (laser)) generate very clear interference strips.

Let us consider examples of interference effects. Think of an oily puddle. The oil creates a thin film on the surface of the water. When a white light shines upon this configuration, the thickness of the oily film will likely coincide with the wavelength of a perceptible color of light. Thus, the reflection at the upper surface of the film and that at the lower surface causes an additive interference effect, allowing the specified color to be observed. Because the oily thickness is not perfectly even across the surface, a rainbow effect is often observed. With multiphase membranes, more complex interference patterns can be produced with various shades of color. The scintillating metallic colors of certain insects are attributed to this effect. The Horyuji temple is painted in this fashion. In South America, the Morpho butterfly shines blue as a result of the lattice structure of its wings. These colors are similarly revealed via the interference of layers (i.e., structural colors).

1.1.4 Polarization

Visible light has the transversal characteristics of all electromagnetic waves. Transverse waves oscillate perpendicularly to their direction of travel. Imagine a beam of light as a 2D construct (e.g., a straight, flat, wave-like banner), imagine having multiple waves of this nature sharing an axis, but each is slightly rotated around the axis. In natural light, this oscillation direction is random. However, the oscillation direction can be aligned using a polarizing filter, such as those attached to the front of a liquid-crystal display (LCD). Light oscillating in a fixed direction is referred to as linearly polarized light. The light that oscillates in a rotating direction in time is referred to as circularly polarized light. Light having both linear and rotation components is referred to as elliptically polarized light. Humans are not sensitive to polarization, but some insects can detect this phenomenon. For example, bees navigate during flight by sensing the polarization of light in the sky.

When non-polarized natural light is reflected by a dielectric surface, such as a plastic or glass surface, the specular component of the reflected light is also polarized, while the diffuse component is not polarized. Light having an oscillation plane orthogonal to the reflection surface is referred to as *p-polarization*, and light having an oscillation plane parallel to the reflection surface is referred to as *s-polarization*. With the specular reflection component, s-polarization is stronger than p-polarization. The incident angle at which the p-polarization becomes zero is called Brewster's angle. Because the ratio of s- to p-polarization depends on the incident angle, this characteristic can be used as a method to determine the incident angle.

1.1.5 Attenuation and Scattering

When light passes through a medium, such as water, it loses energy. This phenomenon is referred to as attenuation. Note that the attenuation described here is different from the phenomenon in which the energy received per unit area decreases as the distance from the light point source increases. Our attenuation is caused by the phenomenon of the energy of the light beam being reduced because of absorption by the intermediate medium. Generally, the energy decreases exponentially along the traveling distance. This can be modeled using the Beer-Lambert law:

$$I_1 = \exp^{-\alpha d} I_0, \tag{1.6}$$

where I_0 and I_1 indicate the original brightness and the brightness after passing the medium of the light, respectively. d is the traveling distance of the light, and α is the parameter determined by the characteristics of the medium. The attenuation of light is particularly noticeable in the ocean; when a diver descends, he/she observes that the brightness of the sunlight decreases until he/she is eventually immersed in darkness. The attenuation of sunlight inside a car is achieved using lead-glass windows. The attenuation of X-rays by lead is also a familiar phenomenon.

When light passes through the air, its traveling direction changes because of the constituent water droplets and particular contaminants of the air. This phenomenon is referred to as scattering. The degree of scattering is related to the size of the matter and the wavelength of the light. Particles and molecules smaller than light wavelengths cause Rayleigh scattering [2], which is more effective at shorter wavelengths and less effective at longer ones. Particles with sizes comparable to a specific wavelength cause Mie scattering [3]. The effect of Mie scattering is independent of wavelength, causing a general white glow. Particles much larger than the wavelength do not cause scattering. They instead cause the shadows and refractions described earlier.

Our sun produces light that is almost white. Thus, for the sake of the current argument, it has evenly distributed energies over all wavelengths. When it arrives at the atmosphere of the earth, scattering occurs because of the atmospheric constituents. Most of these constituents are shorter than visible light wavelengths. Thus, the Rayleigh scattering predominantly produces a blue glow that is observable in all directions.

The non-blue visible wavelengths proceed along their paths without much disturbance. Therefore, when the sun is low in the sky, most of the blue colors are scattered and absorbed before the light reaches us. That is why sunrises and sunsets are mostly red. Alternatively, the contents of clouds are predominately water droplets of varying

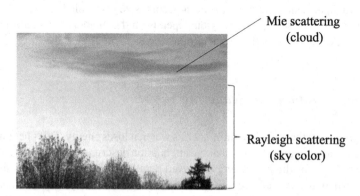

Mie scattering
(cloud)

Rayleigh scattering
(sky color)

Fig. 1.4 Rayleigh scattering and Mie scattering observed in the beautiful sunset-sky of Lake Washington

sizes. Thus, the collective effect observed through clouds is that of Mie scattering. Because the degree of scattering is independent of wavelength, clouds are observed as white (Fig. 1.4).

1.2 Brightness and Physics

Active-lighting methods for computer vision illuminate a target scene using light sources for measurement, reconstruction, and recognition. For this purpose, it is necessary to define the method used to measure the brightness of both the source light and the return light. This section explains the definition of brightness more precisely.

1.2.1 Solid Angle

We use a solid angle to measure the brightness of the light. Before explaining a solid angle, we review the definition of a 2D angle. A 2D angle measures the width of a corner formed by two rays starting from the corner point. We use radian ([rad]) as the unit of measure.

Two rays traveling from the center point of a circle cut out an arc along the circle's circumference. The arc length, α, between the two rays, divided by the radius of the circle, r, is equal to the 2D angle between the two rays, θ

$$\theta = \frac{\alpha}{r} \text{ [rad]}. \tag{1.7}$$

The 2D angle of the entire direction along the whole circle is 2π [rad].

A 3D solid angle is defined as the measurement of a cone's width in 3D. Here, instead of two rays, a set of infinite rays cut out a cone whose vertex is at the center of a sphere to its spherical surface. The directional coverage is the definition of the solid angle. Similar to 2D methods, we can draw a sphere whose center is located at the vertex of the cone. The cone cuts out an area of the spherical surface. The solid angle, Θ, described in steradian units ([sr]), is equal to the circumscribed surface area, A, divided by the square of the spherical radius, r

$$\Theta = \frac{A}{r^2} \text{ [sr]}. \tag{1.8}$$

The 3D solid angle of the entire direction along the whole spherical surface is 4π [sr] (Fig. 1.5).

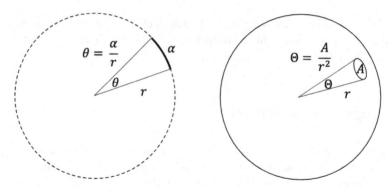

Fig. 1.5 Angle and solid angle

1.2.2 Brightness in Radiometry

In the "Radiometry" field, the brightness of a light source or a surface is an absolute physical quantity without any normalization. However, in the "photometry" field, human sensitivity at each wavelength must be considered. This book measures brightness using various sensing devices, and mainly utilizes radiometric definitions.

A point light source emits light energy as a *radiant flux*, which is measured using [W] (watt) as the unit. One method used to define the brightness of a point light source is measuring how much energy quantity is projected per unit solid angle. This is referred to as radiant intensity, using [W sr^{-1}] as the unit of measure. In some publications, the word "intensity" is used to express the brightness of a light source or a surface without providing a clear definition. Instead, this book uses "brightness" for general cases to avoid confusion.

A surface patch receives light energy from various directions. By integrating the receiving energy over a unit area, we can define how much energy quantity is received by unit area as *irradiance* of the surface. The unit of irradiance is [W/m^2].

When a surface patch is illuminated, the patch looks bright. This is because the surface patch reflects back the light energy. The amount the patch return is defined by *radiance*. To measure the quantity of returning energy, we prepare unit sphere to measure around the patch. The energy from the patch is measured by watt per unit solid angle [W sr^{-1}]. Furthermore, as the surface becomes wider, the more energy is returned. Thus, we must normalize the surface area foreshortened to one observation direction. The unit of the radiance is [W/sr $\cdot$ m^2] (Fig. 1.6).

1.2.3 Brightness to Human

Humans can only see visible light, and the sensitivity differs across wavelengths. For example, when a light source emits a large quantity of energy in the infrared region,

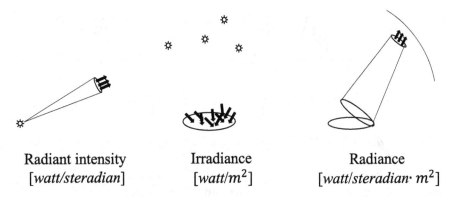

Radiant intensity	Irradiance	Radiance
[*watt/steradian*]	[*watt/m²*]	[*watt/steradian· m²*]

Fig. 1.6 Radiant intensity, irradiance, and radiance

it is not perceived as bright for humans. In practice, since we use light sources to illuminate our surroundings, it is necessary to define brightness according to the human sensitivity. The brightness of a light source, which is weighted to human sensitivities at each wavelength, is referred to as *Luminous flux*, corresponding to radiant flux in the radiometry area. The unit of luminous flux is [lm] (lumen).

Luminous flux per unit solid angles, corresponding to radiant intensity in the radiometry area, denotes how bright the point light source is to humans. This is referred to as *luminous intensity* and is measured by the unit [lm sr^{-1}] or [cd] (candela). Illuminance, corresponding to radiance, is luminous flux incident to a surface and is measured with the unit [lm/m²]. Luminance defines how bright the surface appears to a human and corresponds to radiance, with [lm/sr · m²] as the unit of measure.

1.3 Light and Reflectance

This section discusses the brightness relations by using the two radiometric brightness units, radiance and irradiance. The relationships include the distance from a light source to the object surface, the illuminated brightness versus reflected brightness, and the distance from an object patch to the projected image patch.

1.3.1 Light Source Brightness and Surface Brightness

Let us consider the irradiance, E [W/m²], of an object surface received from a point light source of radiant intensity, I [W sr^{-1}]. See Fig. 1.7a. From the light source, the surface patch spans the spherical angle, $d\omega_o$ [sr]:

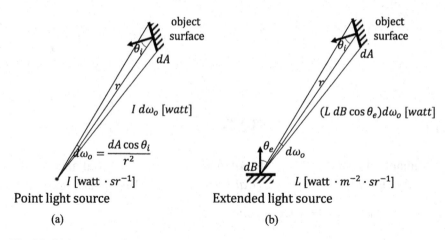

Fig. 1.7 Light source and received energy

$$dw_o = \frac{dA \cos \theta_i}{r^2} \quad [sr], \tag{1.9}$$

where r is the distance between the light source and the surface patch and θ_i is the incident angle of the light source with respect to the normal of the surface patch dA. The energy received by the surface patch dA is $I_i dw_o$ [W]. The irradiance of the surface patch, E, is obtained by dividing the energy by the patch size, dA,

$$E = \frac{I_i dw_o}{dA}$$
$$= I_i \frac{1}{r^2} \cos \theta_i \quad [\text{W/m}^2]. \tag{1.10}$$

The irradiance on the patch, E, is proportional to the radiant intensity of the light source, I_i. However, the farther away the light source, the weaker the receiving irradiance. This matches our common experience in that distant appear look dark. Additionally, when the surface turns away from the light source, it receives less energy.

Next, let us consider the extended light source with the radiance, L, the size, dB, and the emitting angle, θ_e. See Fig. 1.7b. From this extended light source, the light energy, $L dB \cos \theta_e dw_o$ [W], is projected toward the direction of the surface patch, where the spherical angle of the object surface is $dw_o = \frac{dA \cos \theta_i}{r^2}$. By dividing the surface area of the patch, dA, the irradiance of the patch, E, is given as

$$E = (L\,dB\cos\theta_e)\,d\omega_o \left(\frac{1}{dA}\right)$$

$$= L\left(\frac{dB\cos\theta_o}{r^2}\right)\cos\theta_i$$

$$= L\,d\omega_i\cos\theta_i \quad [\text{W/m}^2]. \tag{1.11}$$

where $d\omega_i = \frac{dB\cos\theta_e}{r^2}$ is the spherical angle spanned by the light source observed from the surface patch, dA. This equation states that the receiving irradiance, E, is proportional to the light source radiance, L. For extended light sources, under per unit spherical angle, $d\omega_i = 1$, the irradiance received by the patch, dA, becomes a constant, L, independent of the distance, r. In other word, an extended light source appears to be a constant brightness independent of distance. This is because the farther away the light source, the wider the area of the light source is observed within unit spherical angle.

1.3.2 Bi-Directional Reflectance Distribution Function (BRDF)

When a surface patch receives light energy from a light source, some of the light is reflected. The ratio between incoming irradiance and out-going radiance is defined as Bi-directional reflectance distribution function (BRDF) [4, 5]. See Fig. 1.8.

$$f(\theta_i, \phi_i, \theta_e, \phi_e) = \frac{L_e(\theta_i, \phi_i, \theta_e, \phi_e)}{E_i(\theta_i, \phi_i)} \quad [\text{sr}^{-1}]. \tag{1.12}$$

The BRDF differs with various surfaces. For example, a Lambertian surface is represented as

$$f_{lambert} = \frac{1}{\pi}. \tag{1.13}$$

Therefore, when a light source of radiance L_i exists in the direction of (θ_i, ϕ_i), the surface patch receives the irradiance, $E_i = L_i d\omega_i \cos\theta_i$, obtained from Eq. (1.11). Then, the radiance of the surface $L_{lambert}$ is

$$L_{lambert} = \frac{1}{\pi} L_i d\omega_i \cos\theta_i. \tag{1.14}$$

The reflected radiance of the surface patch depends only on the cosine of the incident angle, not on the emitting angle. This is Lambert's cosine law.

The BRDF of a specular patch is

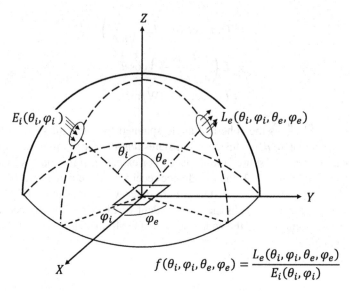

$$f(\theta_i, \varphi_i, \theta_e, \varphi_e) = \frac{L_e(\theta_i, \varphi_i, \theta_e, \varphi_e)}{E_i(\theta_i, \varphi_i)}$$

Fig. 1.8 Bi-directional reflectance distribution function

$$f_{specular} = \frac{\delta(\theta_e - \theta_i)\delta(\phi_e - \phi_i - \pi)}{\sin\theta_i \cos\theta_i}. \tag{1.15}$$

Here $\sin\theta_i$ and $\cos\theta_i$ are the terms required for the result of the spatial integration of δ function to become 1.

$$L_e(\theta_e, \phi_e) = \iint f_{specular} L_i(\theta_i, \phi_i) \sin\theta_i \cos\theta_i d\theta_i d\phi_i$$
$$= L_i(\theta_e, \phi_e - \pi). \tag{1.16}$$

Notably, the out-going radiance is equal to the incoming radiance.

1.3.3 Object Brightness and Image Brightness

An object patch, dA, is projected as an image patch, dI, on the image plane through a lens. Let us suppose that the distance between the surface patch and the lens, the distance between the lens and the image plane, and the diameter of the lens are z, f and d, respectively. Additionally, the angle between the normal at the surface patch, dA, and the straight line, AC, and the angle between the optical axis, the straight line perpendicular to the lens, and the straight line, AC, are θ and α, respectively. See Fig. 1.9.

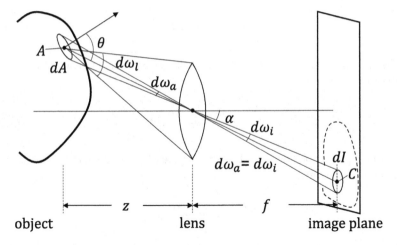

Fig. 1.9 Object and image patch

In the first step, let us consider the light energy going to the lens surface. The radiance of the patch, dA, is L [W/m$^2 \cdot$ sr]. The area of the lens is $\pi(\frac{d}{2})^2 \cos \alpha$. The distance between the surface patch, dA, and the lens is $\frac{z}{\cos \alpha}$ and, thus, the spherical angle of the lens, $d\omega_l$, with respect to the surface patch, dA, is

$$dω_l = \pi \left(\frac{d}{2}\right)^2 \frac{\cos^3 \alpha}{z^2} \quad [\text{sr}]. \tag{1.17}$$

Thus, the incoming light energy to the lens, Φ, is the product of the radiance and the surface patch area and the spherical angle of the lens

$$\begin{aligned}
\Phi &= L(dA \cos \theta)d\omega_l \\
&= L\pi \left(\frac{d}{2}\right)^2 \frac{\cos^3 \alpha}{z^2} dA \cos \theta \quad [\text{W}].
\end{aligned} \tag{1.18}$$

This light energy, Φ, is projected to the patch on the image plane, dI. Thus, the irradiance at the patch, E, is

$$\begin{aligned}
E &= \frac{\Phi}{dI} \\
&= L\pi \left(\frac{d}{2}\right)^2 \frac{\cos^3 \alpha}{z^2} \cos \theta \frac{dA}{dI} \quad [\text{W/m}^2].
\end{aligned} \tag{1.19}$$

The ratio between dI and dA can be obtained when considering the fact that the spherical angle, $d\omega_i$, is equal to the spherical angle, $d\omega_a$, because of the projection relation [6]. First let us consider the spherical angle, $d\omega_i$, which can be obtained from the foreshortening area, $dI \cos \alpha$, and the distance, $\frac{f}{\cos \alpha}$.

$$dw_i = \frac{\cos^3 \alpha}{f^2} dI \ .$$
(1.20)

The spherical angle, dw_a, is obtained as the foreshortening area, $dA \cos \theta$, and the distance, $\frac{z}{\cos \alpha}$

$$dw_a = \frac{\cos \theta \cos^2 \alpha}{z^2} dA \ .$$
(1.21)

From the projection law, these two spherical angles are equal; $dw_i = dw_a$. Thus, we can obtain the ratio,

$$\frac{dA}{dI} = \frac{z^2 \cos \alpha}{f^2 \cos \theta}$$
(1.22)

By substituting this term to the Eq. (1.19), we can obtain the ratio

$$E = L \frac{\pi}{4} \left(\frac{d}{f} \right)^2 \cos^4 \alpha.$$
(1.23)

This equation tells us that the irradiance on the image plane, E, is proportional to the object radiance, L. It also shows that, if the radiance of the object surface is constant, then the irradiance of the projected patch becomes darker as it goes to the peripheral area at the ratio, $\cos^4 \alpha$. The ratio, $\left(\frac{f}{d} \right)$, is referred to as the *f-number* of the lens; and the smaller this value, the brighter the image.

1.3.4 Interreflection

Generally, a point in a scene is not only illuminated by light coming directly from a light source (e.g., LED, bulb, laser). It is also indirectly illuminated by reflected light, as shown in Fig. 1.10. This effect is called *interreflection* and is often non-negligible.

In some studies on active-lighting techniques, interreflection was modeled as light transport, where the effects of indirect light between points in a scene were represented as linear equations. In this section, light transport and its applications related to active-lighting techniques are briefly explained.

Let us consider a case in which a scene is illuminated by an arbitrary light environment, and a camera captures the scene. A light ray captured by the camera may have bounced multiple times before it reached the camera pixel (Fig. 1.10). This situation is often modeled using a light transport equation. A light transport equation is also known as a rendering equation. It is well-known to the computer graphics community, and it describes the equilibrium distribution of radiance in the scene.

Fig. 1.10 Interreflection

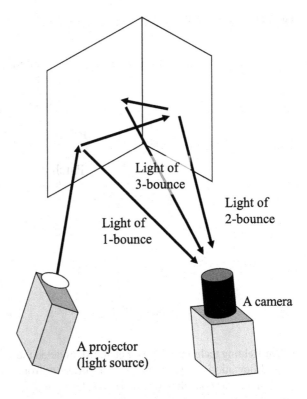

Light of
3-bounce

Light of
2-bounce

Light of
1-bounce

A camera

A projector
(light source)

Consider a fixed surface point, $\mathbf{x}$, and a camera position, $\mathbf{c}$. Let us denote a surface point around $\mathbf{x}$ that illuminates $\mathbf{x}$ as $\mathbf{y}$, the direction from $\mathbf{x}$ to $\mathbf{c}$ as $\omega_{\mathbf{x}\to\mathbf{c}}$, and the direction from $\mathbf{y}$ to $\mathbf{x}$ as $\omega_{\mathbf{y}\to\mathbf{x}}$. The camera can observe the radiance of $\mathbf{x}$ to the direction, $\omega_{\mathbf{x}\to\mathbf{c}}$, which we denote as $l_{out}(\mathbf{x}, \omega_{\mathbf{x}\to\mathbf{c}})$.

All $l_{out}(\mathbf{x}, \omega_{\mathbf{x}\to\mathbf{c}})$ of point $\mathbf{x}$ can be divided into two components of illumination: direct and indirect. The direct component is the radiance caused by the reflected light from light sources that is not included in the scene. The indirect component is caused by reflecting light that comes from other points, $\mathbf{y}$, in the scene.

Let us denote the direct component of $l_{out}(\mathbf{x}, \omega_{\mathbf{x}\to\mathbf{c}})$ as $l_{direct}(\mathbf{x}, \omega_{\mathbf{x}\to\mathbf{c}})$. Then, using the from of the light transport equation, $l_{out}(\mathbf{x}, \omega_{\mathbf{x}\to\mathbf{c}})$ can be written as follows: [7]:

$$l_{out}(\mathbf{x}, \omega_{\mathbf{x}\to\mathbf{c}}) = l_{direct}(\mathbf{x}, \omega_{\mathbf{x}\to\mathbf{c}}) + \int_{\mathbf{y}} A(\omega_{\mathbf{y}\to\mathbf{x}}, \omega_{\mathbf{x}\to\mathbf{c}}) l_{out}(\mathbf{y}, \omega_{\mathbf{y}\to\mathbf{x}}) \, d\mathbf{y}, \quad (1.24)$$

where $A(\omega_{\mathbf{y}\to\mathbf{x}}, \omega_{\mathbf{x}\to\mathbf{c}})$ denotes the proportion of irradiance from point $\mathbf{y}$ to $\mathbf{x}$ that gets transported as radiance in the direction of $\mathbf{c}$, and the integral variable, $\mathbf{y}$, traverses all surface points of the scene. $A(\omega_{\mathbf{y}\to\mathbf{x}}, \omega_{\mathbf{x}\to\mathbf{c}})$ includes geometrical properties, such as the distance and occlusion between $\mathbf{x}$ and $\mathbf{y}$. This process of interreflection causes considerable effects on the results of photometric stereo, which is one of major

Fig. 1.11 Light transport

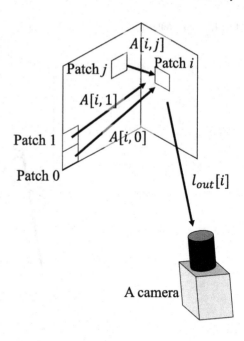

active-lighting techniques. Nayar et al. [8] modeled the process where interreflection caused deviations of photometric stereo results from the standard case (i.e., only direct light was considered) and proposed a method for canceling the interreflection effect by iteratively applying the inverse process.

Equation (1.24) can be rewritten as the discretized formulation by dividing the scene surfaces into small patches as shown in Fig. 1.11, and replacing the integration with summation

$$l_{out}[i] = l_{direct}[i] + \sum_j A[i, j] l_{out}[j]. \tag{1.25}$$

This can be rewritten as a matrix equation

$$\mathbf{l}_{out} = \mathbf{l}_{direct} + \mathbf{A} \mathbf{l}_{out}. \tag{1.26}$$

Thus

$$\mathbf{l}_{out} = (\mathbf{I} - \mathbf{A})^{-1} \mathbf{l}_{direct}. \tag{1.27}$$

Seitz et al. [9] showed that there exists a linear operator, $\mathbf{C}^1$, that separates the direct component (i.e., first-bounce component) from the output radiances of the scene, because, from Eq. (1.27),

$$\mathbf{l}_{direct} = (\mathbf{I} - \mathbf{A})\mathbf{l}_{out} \equiv \mathbf{C}^1 \mathbf{l}_{out}. \tag{1.28}$$

Seitz et al. also showed that the k-th bounce component, which bounces exactly k times in the scene, can be separated by iteratively applying the same operator, $\mathbf{C}^1$. They captured a large number of images while moving a computer-controlled pan-tilt laser beam as a light source. They calculated the canceling operator from the images, and separated the brightness values into images having k-bounced lights for $k = 1, 2, \cdots$. They further observed that the concave part of the scenes was strongly illuminated from the indirect lights.

Seitz et al. used a large number of images with laser-bean projection to measure the light transport of a scene. Nayar et al. [10] showed that the direct and indirect components of the light transport could be separated by capturing a small number of images via structured light projection.

Nayar et al. considered a setting wherein a scene was illuminated by a projector and captured by a camera. From Eq. (1.25), Nayar et al. showed that the direct component of a patch the scene depended on a small number of projector pixels that directly illuminated the patch, whereas the indirect component depended upon a large number of pixels, because they illuminated the patch indirectly via single or multiple reflections.

Therefore, the light transport function of direct components from the projector to the camera becomes a function with the support of a small number of pixels (close to a delta function). However, the light transport of the indirect (global) component becomes a function with the support of many pixels. The shape of the transport function of indirect components is complicated and depends on many factors, such as the scene shape, camera and projector arrangements, BRDFs, etc. However, it is generally true that, compared with the number of projector pixels generating direct components, the number of projector pixels generating global luminance components is overwhelming. Thus, the indirect component values are blurred because of its larger support, causing a smoothed light transport function.

In summary, the transport function of the direct component has high spatial frequencies with respect to the projector pixels, and the indirect (global) component becomes a function having low spatial frequencies. Using the differences of the spacial frequency properties between the direct and indirect components, they showed that those components could be separated from at least two images.

1.4 Radiometric Calibration

Radiometric calibration ensures the proper measurement of electromagnetic radiation. In the contexts of active-lighting and computer vision methods, there are two important radiometric calibrations. The first is *sensor calibration*, which determines how the incoming light is mapped to the measured pixel brightness. The other is *light source calibration*, which is concerned with the angular distribution of light emissions. This section briefly explains the basics of radiometric calibration.

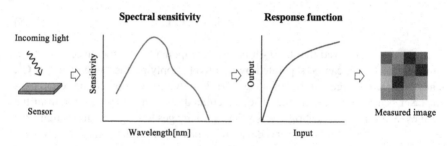

Fig. 1.12 Illustration of the in-camera imaging pipeline

1.4.1 Radiometric Calibration of Image Sensors

Digital image sensors collect incoming light, produce electrons, and convert them into digital signals via analog-to-digital conversion. In this manner, the pixel brightness is related to the strength of the incoming light. Figure 1.12 shows a simplified in-camera imaging pipeline. From the incoming light to the measured pixel brightness conversion, there are two major components wherein the strength of the incoming light is transformed. The first is *spectral sensitivity*, which represents the relative efficiency of detecting the incoming light that depends on the wavelength. A sensor has its own characteristic curve of the responsivity, which describes the photo-current per incident unit optical power. The second component is the *radiometric response function*, which relates the detected signal (light) to a pixel brightness, which is typically nonlinear for various consumer cameras. In this subsection, we describe the radiometric calibration methods for these components.

Calibration of Spectral Sensitivity

Light spans a wide range of wavelengths. With a digital color camera, the scene radiance is recorded as RGB values via color filters that specify light at different wavelength ranges to be observed. Therefore, the recorded RGB values depend on the spectral sensitivity of color filters and sensors, i.e., different sensors yield different RGB outputs for the same scene. Camera spectral sensitivity plays an important role in various active-lighting and computer vision tasks including multi-spectral imaging, color rendering, and color constancy, which rely on measured color information. Therefore, calibration of camera spectral sensitivity is necessary.

The spectral sensor response $r \in \mathbb{R}$ to the wavelength $\lambda \in \mathbb{R}$ in the range ω can be modeled as

$$r = e \int_{\omega} s(\lambda)\rho(\lambda)l(\lambda)d\lambda, \tag{1.29}$$

where e is the exposure time, $s(\lambda)$ is the spectral response function, $\rho(\lambda)$ is the spectral reflectance, and $l(\lambda)$ is the spectral power distribution of the illuminant. The goal of the calibration of spectral sensitivity is to determine the spectral response function $s(\lambda)$.

A standard technique for estimating camera spectral sensitivity $s(\lambda)$ is taking pictures of monochromatic light whose bandwidth is narrow (i.e., the range of the wavelength ω is small) using a monochromator or narrow-band filters. The spectral sensitivity of the camera can then be reliably estimated from recorded observations and known spectral distributions of monochromatic light. By repeating the procedure for various wavelengths of the monochromatic light by fixing the sensor and the target scene, the distribution of the spectral sensitivity $s(\lambda)$ across the wavelength λ can be obtained.

Although the above procedure provides accurate estimates, the method requires expensive hardware to generate and measure the series of monochromatic light. Thus, its use has been limited to well-equipped laboratories only. Additionally, the entire procedure is laborious because of the need for multiple observations at different wavelengths of light. To simplify the procedure, methods using calibration targets whose reflectance is known have been proposed [11–13]. These methods estimate the camera spectral sensitivity by recording a calibration target (e.g.., IT8 Target or Macbeth color checker) that contains several patches whose spectral reflectances are known, under the illumination of a known spectral distribution. These methods require controlled lighting, under which light is made spatially uniform, with a known illumination spectrum.

To ease the requirements of providing a known illumination spectrum, Han et al. proposed a method based on a specialized fluorescent chart [14]. DiCarlo et al. used an emissive calibration chart with LEDs of known spectra [15]. Regarding analysis, there is a study on the space of camera spectral sensitivity [16]. With this study, it is found that the space is convex and two-dimensional. This observation has been shown useful for developing a more advanced spectral sensitivity calibration method.

Calibration of Sensor Response Functions

In most cameras, there exists a *radiometric response function* that relates the incoming light to the sensor to measured pixel brightness values. For purposes of compressing the dynamic range of scene brightnesses or accounting for the nonlinear mapping of display systems, the radiometric response functions are often intentionally designed to nonlinearly.

Although many active-lighting techniques assume a linear (or affine) relationship between the sensor irradiance and the pixel brightness, radiometric response functions are typically unknown and vary with image sensors and camera parameter settings. For those methods that require the linearity of the pixel brightnesses with respect to the strength of incoming light, *radiometric calibration* is needed.

The radiometric response function $f : \mathbb{R} \to \mathbb{R}$ maps the strength of incoming light $r \in \mathbb{R}$ that is recorded at the sensor to the image brightness $m \in \mathbb{R}$ by

$$m = f(r). \tag{1.30}$$

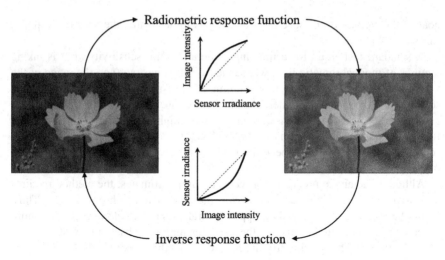

Fig. 1.13 Sensor irradiance is mapped to a measured pixel brightness via a radiometric response function. The inverse mapping is described by its inverse response function

For active-lighting methods that require the information about the strength of incoming light r rather than measured pixel brightnesses m as input, the *inverse response function* $g = f^{-1}$ needs to be determined so that the measured pixel brightness m can be made linear with respect to the incoming light r. Because most of response functions f are continuous and monotonic, they are invertible to yield inverse response functions (see Fig. 1.13).

Representation

There are various parametric expressions for radiometric response functions. To deal with the unknown scaling between the incoming light r and measured pixel brightness m, both are normalized in the range of $[0, 1]$, so that $f(0) = 0$ and $f(1) = 1$. The choice of the parametric representation of radiometric response functions comes with a trade-off of complexity and flexibility. A simpler representation makes the radiometric calibration easier at the cost of approximation accuracy. On the other hand, a more flexible representation requires a difficult solution method. In what follows, we review a few representations that are used in the literature.

Mann and Picard [17] used a gamma correction function to represent radiometric response functions as

$$m = f(r) = \alpha + \beta e^{\gamma}, \tag{1.31}$$

where $\alpha, \beta \in \mathbb{R}$ are offset and scaling factors, and $\gamma \in \mathbb{R}$ is a power-law parameter. Mitsunaga and Nayar [18] used a polynomial function as the model of inverse response functions

$$r = g(m) = \sum_{n=0}^{N} c_n m^n, \tag{1.32}$$

where $\{c_n\}$ are coefficients of the polynomial function. Grossberg and Nayar applied principal component analysis (PCA) to a database of real-world response functions (DoRF) and showed that the space of response functions could be represented by a small number of basis functions [19]:

$$g = \bar{g} + \sum_{n=1}^{N} c_n g_n, \tag{1.33}$$

in which $\bar{g}$ is the mean of all the inverse response functions, and g_i is the i-th principal component of the inverse response functions. Debevec and Malik [20] used a nonparametric form of the radiometric response functions. The nonparametric expression has a great descriptive power, but it requires determining $f(r)$ (or $g(m)$) at each brightness level, e.g.., 256 levels for 8-bit images. The same representation has also been used by Tsin et al. [21].

Methods of Calibrating Sensor Response Functions

Radiometric calibration of sensor response functions aims at estimating the response function f of a camera in Eq. (1.30). For active-lighting methods that require irradiance values r rather than measured pixel brightnesses m as input, the inverse response function $g = f^{-1}$ needs to be determined so that measured pixel brightnesses m can be made linear with respect to the sensor irradiances r. Assuming that the response functions f are monotonic and continuous, there is a one-to-one mapping between the response function f and its inverse response function g.

Radiometric calibration methods require means of collecting samples of the radiometric response function with some known relationship. One of the traditional approaches is using a specialized target, such as a Macbeth color chart [22], which has color patches with known reflectances. By uniformly illuminating the target, radiances from the color patches are recorded. The radiometric response function is then obtained by relating the sensor irradiance r to the recorded pixel brightness values m. Nayar and Mitsunaga [23] used an optical filter with spatially varying transmittance, where the variation corresponded to the radiance ratio.

To mitigate the need for such special equipments, some methods use a set of static-scene images observed from a fixed viewpoint, taken with different and known exposure times, so the radiance ratio can be directly obtained. The known exposure times provide information about sensor irradiance ratios, which are proportional to the exposure times. Similar to using a specialized target, by relating the sensor irradiance r with the measured pixel brightnesses m, a radiometric response function can be estimated. Because it does not require a tailored calibration target, the multi-exposure approach is a veteran but still frequently used method (see Fig. 1.14).

An early work of Mann and Picard [17], uses a gamma correcting function to represent response functions. With known exposure ratios, their method can successfully recover the inverse response function in the parametric form. With only approximate knowledge of relative exposure levels, Mitsunaga and Nayar [18], iteratively solve for a response function based on the assumption that it has a polynomial

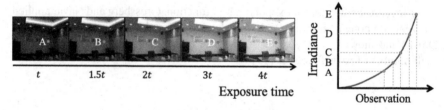

Fig. 1.14 Radiometric calibration of sensor responses by multiple known exposures

form. Other iterative estimation methods include Tsin et al.'s method [21], which estimates nonparametric responses based on a statistical model of the CCD imaging process, and Pal et al.'s method [24], which uses probabilistic imaging models and priors of response functions to determine the response functions that can potentially differ from image to image. Debevec and Malik [20] assumed a smoothness property of the response functions and estimated them in a nonparametric manner. As pointed out in [25–27], without the knowledge of exposure ratios, such an estimate suffers from an *exponential ambiguity*. While not unique, such an estimate is still useful for many applications, such as radiometric alignment, high dynamic range image production, and image stitching.

Several methods have been developed that use multiple exposures but do not require precise image alignment. Grossberg and Nayar [25] use the relationship between the brightness histograms of two scenes imaged with different exposures because the brightness histograms are relatively unaffected by small changes of the scene. Kim and Pollefeys [26] compute point correspondences between images. Mann [28] estimates response functions from a rotating and zooming camera.

Instead of using varying exposure times, some approaches use statistical or physical properties embedded in images to achieve radiometric calibration. Tsin et al.'s method [21] estimates nonparametric response functions using a statistical model of the CCD imaging process. Pal et al. [24] used probabilistic imaging models and weak prior models for deriving response functions to produce high-quality high dynamic range images. Lin et al. [29], and Lin and Zhang [30], proposed a method that takes only a single image as input. Their method uses edges for obtaining color or gray scale histogram distributions, and the optimal inverse response function is determined by transforming linear distributions. Their method uses a database of response functions (DoRF) compiled by Grossberg and Nayar [31]. In a similar manner, Wilburn et al. [32], use temporal color mixtures to directly sample the response function by observing motion blur in an image. More recently, Shi et al. [33], show a calibration method from images taken under varying lighting conditions. In their approach, an inverse response function is determined by linearizing color profiles that are defined as a set of measured RGB values at a pixel across images.

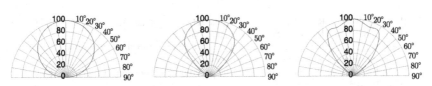

Fig. 1.15 Examples of radiant intensity distributions (RIDs) plotted in polar coordinates. The left-most shows an isotropic RID, whereas the others are anisotropic RIDs

1.4.2 Radiometric Calibration of Light Sources

Calibration of light sources is often an important first step of various active-lighting methods. For example, shape-from-shading [34], photometric stereo [35], and reflectometry [36] methods often require accurate knowledge of light sources for recovering surface normal maps and reflectances. While *geometric light calibration*, such as estimating an incident light direction or the position of near point light source has been well studied [37–40], the radiometric calibration of the light sources has been less studied in the literature.

Accurate radiometric calibration, such as radiant intensity distributions (RIDs) of a point light source is important. The RID may be *isotropic* if the radiant intensity is the same in all directions, or *anisotropic* (or non-isotropic) otherwise. Figure 1.15 shows some of the RIDs plotted in polar coordinates. The calibration of RIDs is usually conducted with a controlled setup using specialized equipments. Indeed, accurate and reliable nonparametric recovery of RIDs of anisotropic point lights currently requires special-purpose equipments, such as an imaging sphere [41] or a goniophotometer [42].

To ease the cost of RID calibration, some studies have investigated more flexible calibration methods. These works used parametric expressions to represent RIDs, turning the calibration task into a parameter estimation problem. For example, Park et al. [43] and Visentini-Scarzanella and Kawasaki [44] used low-order polynomial expressions for RIDs. Another popular expression for RIDs use powers of cosine functions [45]. To extend the representation capability, Moreno and Sun [46] used the sum of a maximum of two or three Gaussian or cosine-power functions.

References

1. Lambert JH (18922) Lamberts Photometrie: Photometria, sive De mensura et gradibus luminus, colorum et umbrae (1760), vol 6. W. Engelmann
2. Strutt JW (1871) Xv. on the light from the sky, its polarization and colour. Lond Edinb. Dublin Philos Mag J Sci 41(271):107–120
3. Horvath H (2009) Gustav mie and the scattering and absorption of light by particles: historic developments and basics. J Quant Spectrosc Radiat Trans 110(11):787–799
4. Nicodemus FE, Richmond JC (1977) Geometrical considerations and nomenclature for reflectance

5. Horn BK, Sjoberg RW (1979) Calculating the reflectance map. Appl Opt 18(11):1770–1779
6. Horn B, Klaus B, Horn P (1986) Robot vision. MIT Press
7. Kajiya JT (1986) The rendering equation. In: Proceedings of the 13th annual conference on computer graphics and interactive techniques, pp 143–150
8. Nayar SK, Ikeuchi K, Kanade T (1991) Shape from interreflections. Int J Comput Vis (IJCV) 6(3):173–195
9. Seitz SM, Matsushita Y, Kutulakos KN (2005) A theory of inverse light transport. In: Proceedings of the international conference on computer vision (ICCV), vol 2. IEEE, pp 1440–1447
10. Nayar SK, Krishnan G, Grossberg MD, Raskar R (2006) Fast separation of direct and global components of a scene using high frequency illumination. In: ACM transactions on graphics (TOG), vol 25. ACM, pp 935–944
11. Finlayson GD, Hordley S, Hubel PM (1998) Recovering device sensitivities with quadratic programming. In: Color and imaging conference, vol 1998. Society for imaging science and technology, pp 90–95
12. Urban P, Desch M, Happel K, Spiehl D (2010) Recovering camera sensitivities using target-based reflectances captured under multiple led-illuminations. In: Proceedings of workshop on color image processing, pp 9–16
13. Rump M, Zinke A, Klein R (2011) Practical spectral characterization of trichromatic cameras. In: ACM transactions on graphics (TOG), vol 30. ACM, p 170
14. Han S, Matsushita Y, Sato I, Okabe T, Sato Y (2012) Camera spectral sensitivity estimation from a single image under unknown illumination by using fluorescence. In: Proceedings of the IEEE conference on computer vision and pattern recognition (CVPR). IEEE, pp 805–812
15. DiCarlo JM, Montgomery GE, Trovinger SW (2004) Emissive chart for imager calibration. In: Color and imaging conference, vol 2004. Society for Imaging Science and Technology, pp 295–301
16. Jiang J, Liu D, Gu J, Süsstrunk S (2013) What is the space of spectral sensitivity functions for digital color cameras? In: 2013 IEEE workshop on applications of computer vision (WACV). IEEE, pp 168–179
17. Mann S, Picard RW (1995) On being 'undigital' with digital cameras: extending dynamic range by combining differently exposed pictures
18. Mitsunaga T, Nayar SK (1999) Radiometric self calibration. In: Proceedings. 1999 IEEE computer society conference on computer vision and pattern recognition (Cat. No PR00149), vol 1. IEEE, pp 374–380
19. Grossberg MD, Nayar SK (2004) Modeling the space of camera response functions. IEEE Trans Pattern Anal Mach Intell (PAMI) 26(10):1272–1282
20. Debevec PE, Malik J (1997) Recovering high dynamic range radiance maps from photographs. In: Proceedings of the 24th annual conference on computer graphics and interactive techniques, pp 369–378
21. Tsin Y, Ramesh V, Kanade T (2001) Statistical calibration of CCD imaging process. In: Proceedings of the international conference on computer vision (ICCV), pp 480–487
22. Chang YC, Reid JF (1996) RGB calibration for color image analysis in machine vision. IEEE Trans Image Process (TIP) 5(10):1414–1422
23. Nayar SK, Mitsunaga T (2000) High dynamic range imaging: spatially varying pixel exposures. In: Proceedings of the IEEE conference on computer vision and pattern recognition (CVPR), vol 1. IEEE, pp 472–479
24. Pal C, Szeliski R, Uyttendale M, Jojic N (2004) Probability models for high dynamic range imaging. In: Proceedings of the IEEE conference on computer vision and pattern recognition (CVPR), pp 173–180
25. Grossberg M, Nayar S (2003) Determining the camera response from images: What is knowable? IEEE Trans Pattern Anal Mach Intell (PAMI) 25(11):1455–1467
26. Kim SJ, Pollefeys M (2004) Radiometric alignment of image sequences. In: Proceedings of the IEEE conference on computer vision and pattern recognition (CVPR), pp 645–651
27. Litvinov A, Schechner YY (2005) Addressing radiometric nonidealities: a unified framework. In: Proceedings of the IEEE conference on computer vision and pattern recognition (CVPR), pp 52–59

28. Mann S (2001) Comparametric imaging: estimating both the unknown response and the unknown set of exposures in a plurality of differently exposed images. In: Proceedings of the IEEE conference on computer vision and pattern recognition (CVPR), pp 842–849

29. Lin S, Gu J, Yamazaki S, Shum HY (2004) Radiometric calibration from a single image. In: Proceedings of the IEEE conference on computer vision and pattern recognition (CVPR), pp 938–945

30. Lin S, Zhang L (2005) Determining the radiometric response function from a single grayscale image. In: Proceedings of the IEEE conference on computer vision and pattern recognition (CVPR), pp 66–73

31. Grossberg M, Nayar S (2003) What is the space of camera response functions? In: Proceedings of the IEEE conference on computer vision and pattern recognition (CVPR), pp 602–609

32. Wilburn B, Xu H, Matsushita Y (2008) Radiometric calibration using temporal irradiance mixtures. In: Proceedings of the IEEE conference on computer vision and pattern recognition (CVPR)

33. Shi B, Matsushita Y, Wei Y, Xu C, Tan P (2010) Self-calibrating photometric stereo. In: Proceedings of the IEEE conference on computer vision and pattern recognition (CVPR), pp 1118–1125

34. Horn BK (1975) Obtaining shape from shading information. The psychology of computer vision, pp 115–155

35. Woodham RJ (1980) Photometric method for determining surface orientation from multiple images. Opti Eng 19(1):191139

36. Romeiro F, Vasilyev Y, Zickler T (2008) Passive reflectometry. In: Proceedings of the European conference on computer vision (ECCV). Springer, pp 859–872

37. Wong KYK, Schnieders D, Li S (2008) Recovering light directions and camera poses from a single sphere. In: Proceedings of the European conference on computer vision (ECCV). Springer, pp 631–642

38. Ackermann J, Fuhrmann S, Goesele M (2013) Geometric point light source calibration. In: Vision, modeling and visualization, pp 161–168

39. Kato K, Sakaue F, Sato J (2010) Extended multiple view geometry for lights and cameras from photometric and geometric constraints. In: Proceedings of the international conference on pattern recognition (ICPR). IEEE, pp 2110–2113

40. Santo H, Waechter M, Samejima M, Sugano Y, Matsushita Y (2018) Light structure from pin motion: simple and accurate point light calibration for physics-based modeling. In: Proceedings of the European conference on computer vision (ECCV), pp 3–18

41. Rykowski R, Kostal H (2008) Novel approach for led luminous intensity measurement. In: Light-emitting diodes: research, manufacturing, and applications XII, vol 6910. International Society for Optics and Photonics, p 69100C

42. Simons RH, Bean AR (2008) Lighting engineering: applied calculations. Routledge

43. Park J, Sinha SN, Matsushita Y, Tai Y, Kweon I (2014) Calibrating a non-isotropic near point light source using a plane. In: Proceedings of the IEEE conference on computer vision and pattern recognition (CVPR), pp 2267–2274. https://doi.org/10.1109/CVPR.2014.290

44. Scarzanella MV, Kawasaki H (2015) Simultaneous camera, light position and radiant intensity distribution calibration. In: Pacific-rim symposium on image and video technology (PSVIT), pp 557–571. https://doi.org/10.1007/978-3-319-29451-3_44

45. Gardner IC (1947) Validity of the cosine-fourth-power law of illumination. J Res Natl Bur Stand 39(3):213–219

46. Moreno I, Sun CC (2008) Modeling the radiation pattern of leds. Opt Express 16(3):1808–1819

Chapter 2
Geometry

A common objective for using active-lighting techniques is to measure 3D geometric properties (e.g., shapes, positions, normals) of real scenes. Thus, 3D geometrical modeling of 3D scenes and measurement devices (e.g., cameras, laser sources, video projectors) are crucially important.

This chapter provides the necessary basic knowledge of the geometric properties of cameras and various light sources. This includes the mathematical formulas of device properties and the basic methods of estimating formula parameters. Regarding light sources, we focus on formulations of point lasers, line lasers, and video projectors, because they are the most commonly used devices for active-stereo methods, which entail the most geometrical active-lighting approaches.

2.1 Camera Model

Camera model is a central topic of computer vision research. For 3D reconstruction techniques, it is critical, because it portrays the 3D world using 2D images. In this section, various, practical camera models are described.

2.1.1 Perspective Projection and Pinhole Cameras

Formally, a camera model is a calculation form of projection in which a point in 3D space is mapped to a point on a 2D image plane. Normally, for computer vision, perspective projection is used. The most typical and the simplest camera model of perspective projection is that of a pinhole camera.

© Springer Nature Switzerland AG 2020
K. Ikeuchi et al., *Active Lighting and Its Application for Computer Vision*,
Advances in Computer Vision and Pattern Recognition,
https://doi.org/10.1007/978-3-030-56577-0_2

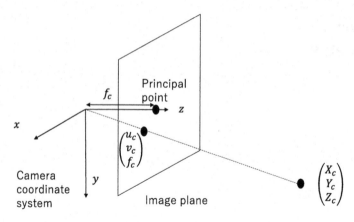

Fig. 2.1 Perspective projection

The geometry of perspective projection is shown in Fig. 2.1. Points in 3D space are represented using a camera coordinate system. Figure 2.1 shows the default camera coordinate system of the OpenCV library [1], where the x-axis is directed toward the right of the camera, the y-axis is directed toward the bottom, and the z-axis is directed toward the front. This 3D coordinate system is a right-handed coordinate system. Different computer vision and graphics libraries are known to use different default coordinate systems. Thus, the users of those library packages should take note of the coordinate systems.

The points of an image plane are represented using 2D coordinates defined by u and v axes. For a normalized camera, whose formulation will be described in Sect. 2.1.4, the origin of the 2D coordinate system is at the cross point of the image plane and the z-axis. This point is called the principal point. The directions of u- and v-axes are defined as directions the same as the x- and y-axes, respectively.

In Fig 2.1 the image plane is placed in front of the camera at $z = f_c$, where f_c is known as the focal length of the camera. In the formulation of perspective projection model, f_c can be regarded as the magnification ratio of the image on the image plane with respect to the 3D space. Let the coordinates of a 3D point be (X_c, Y_c, Z_c), and let the coordinates of the corresponding 2D point be (u_c, v_c). Then, the equations of perspective projection are

$$u_c = \frac{f_c X_c}{Z_c}$$
$$v_c = \frac{f_c Y_c}{Z_c}. \tag{2.1}$$

For a digital camera, the image plane consists of an optical imaging device, such as a CCD, instead of the film used with analog cameras. Thus, the 2D signal on the image plane is detected as a digital image that is sampled on the device's pixels. Normally, 2D coordinates on the digital image are represented as coordinates, in which the size of the pixel intervals is the unit length. Its origin is located at the top left of the digital image in the case of the OpenCV library. Thus, the 2D coordinate system of the digital image is different from the 2D coordinate system of the normalized camera. Let the location of the principal point be (c_x, c_y) using the coordinates of the digital image and let the sizes of the pixel intervals on the image plane of the normalized camera be Δ_u and Δ_v for the u and v directions, respectively. Then

$$u = \frac{f_x X_c}{Z_c} + c_x$$
$$v = \frac{f_y Y_c}{Z_c} + c_y, \qquad (2.2)$$

where (u, v) is the 2D coordinate of the digital image, $f_x \equiv \frac{f_c}{\Delta_u}$, and $f_y \equiv \frac{f_c}{\Delta_v}$.

2.1.2 Coordinate Transformation

To model the position or the motion of the camera in 3D space, or to model multiple cameras, multiple coordinate systems corresponding to each camera position for each camera device or frame are used. The geometric relationships between these coordinates are represented as rigid transformations.

Assume that we have a camera, 1, in a world coordinate system, as shown in Fig. 2.2. Its position is assumed to be general. Let a point represented using camera coordinates of camera 1 be (X_c^1, Y_c^1, Z_c^1). The same point is represented in world coordinates as (X_w, Y_w, Z_w). Then, the transformation between these coordinates is depicted as

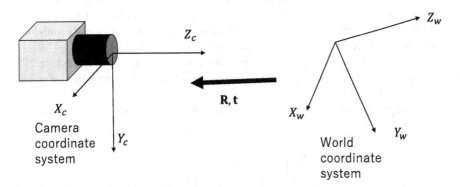

Fig. 2.2 Coordinate transformation

$$\begin{bmatrix} X_c^1 \\ Y_c^1 \\ Z_c^1 \end{bmatrix} = \mathbf{R} \begin{bmatrix} X_w \\ Y_w \\ Z_w \end{bmatrix} + \mathbf{t}, \tag{2.3}$$

where $\mathbf{R}$ is a rotation matrix, and $\mathbf{t}$ is a translation vector.

2.1.3 Homogeneous Coordinates

Mappings of perspective projections and transformations between different coordinate systems can be simply represented using homogeneous coordinates. A point in 3D space is represented by four homogeneous coordinate elements, whereas the same point is represented by three in a normal coordinate system (i.e., Euclidean coordinates). $[a, b, c, d]^\top$, where $d \neq 0$ in homogeneous coordinates, represents $[a/d, b/d, c/d]^\top$ in Euclidean coordinates. $[a, b, c, 0]^\top$ in homogeneous coordinates represents the point at infinity in the direction of $\pm[a, b, c]^\top$ in Euclidean coordinates. Note that the point at infinity along the direction of $+[a, b, c]^\top$ and that of $-[a, b, c]^\top$ are treated to be identical points. Inversely, a point represented by $[X, Y, Z]^\top$ in Euclidean 3D coordinates can be represented by $s[X, Y, Z, 1]^\top$, where $s \neq 0$ in homogeneous coordinates.

Similarly, a point represented by $[u, v]^\top$ in Euclidean 2D coordinates is represented as $s[u, v, 1]^\top$, where $s \neq 0$ in homogeneous coordinates.

A mapping between points calculated using matrix multiplication between their homogeneous coordinate representations is a projective transformation. This type of transformations plays an important role in computer vision and computer graphics. Let a point represented by $\tilde{\mathbf{x}}_1 = [a_1, b_1, c_1, d_1]^\top$ in homogeneous coordinates be transformed into a point, $\tilde{\mathbf{x}}_2 = [a_2, b_2, c_2, d_2]^\top$, in homogeneous coordinates, where $\tilde{\mathbf{x}}_2$ is calculated by $\tilde{\mathbf{x}}_2 = \mathbf{A}\tilde{\mathbf{x}}_1$ and $\mathbf{A}$ is a 4×4 matrix. Because homogeneous coordinates have scale ambiguity (i.e., $\tilde{\mathbf{x}}_2$ and $s\tilde{\mathbf{x}}_2$ represent the same point if $s \neq 0$), $\mathbf{A}$ and $s\mathbf{A}$, ($s \neq 0$) represent the same transformation. Generally, a projective transformation represented by a 4×4 matrix, $\mathbf{A}$, has $4 \times 4 - 1 = 15$ DoF.

Similarly, a projective transformation from a 3D point, $\tilde{\mathbf{x}}$, in 3D space to a 2D point, $\tilde{\mathbf{u}}$, in 2D space can be represented by a 3×4 matrix, $\mathbf{P}$. This transformation has $3 \times 4 - 1 = 11$ DoF.

By using homogeneous coordinates and projective transformations, Eq. (2.2), can be written as

$$s \begin{bmatrix} u \\ v \\ 1 \end{bmatrix} = \begin{bmatrix} f_x & 0 & c_x & 0 \\ 0 & f_y & c_y & 0 \\ 0 & 0 & 1 & 0 \end{bmatrix} \begin{bmatrix} X_c \\ Y_c \\ Z_c \\ 1 \end{bmatrix}. \tag{2.4}$$

Similarly, a rigid transformation of Eq. (2.3), can be written as

$$
\begin{bmatrix} X_c^1 \\ Y_c^1 \\ Z_c^1 \\ 1 \end{bmatrix} = \begin{bmatrix} \mathbf{R} & \mathbf{t} \\ \mathbf{0}^\top & 1 \end{bmatrix} \begin{bmatrix} X_w \\ Y_w \\ Z_w \\ 1 \end{bmatrix}. \tag{2.5}
$$

By combining the above two equations, the projection of a point represented by (X_w, Y_w, Z_w) in an arbitrary coordinate system (e.g., the world coordinates) into a point in the digital image coordinates, (u, v), can be represented by

$$
s \begin{bmatrix} u \\ v \\ 1 \end{bmatrix} = \begin{bmatrix} f_x & 0 & c_x & 0 \\ 0 & f_y & c_y & 0 \\ 0 & 0 & 1 & 0 \end{bmatrix} \begin{bmatrix} \mathbf{R} & \mathbf{t} \\ \mathbf{0}^\top & 1 \end{bmatrix} \begin{bmatrix} X_w \\ Y_w \\ Z_w \\ 1 \end{bmatrix} = \begin{bmatrix} f_x & 0 & c_x \\ 0 & f_y & c_y \\ 0 & 0 & 1 \end{bmatrix} [\mathbf{R} \ \mathbf{t}] \begin{bmatrix} X_w \\ Y_w \\ Z_w \\ 1 \end{bmatrix}. \tag{2.6}
$$

More generally, the following equation obtained by adding a *skew* parameter, k, into Eq. (2.6) is often used

$$
s \begin{bmatrix} u \\ v \\ 1 \end{bmatrix} = \mathbf{K} \begin{bmatrix} X_c \\ Y_c \\ Z_c \end{bmatrix} = \mathbf{K} [\mathbf{R} \ \mathbf{t}] \begin{bmatrix} X_w \\ Y_w \\ Z_w \\ 1 \end{bmatrix},
$$

$$
\mathbf{K} \equiv \begin{bmatrix} f_x & k & c_x \\ 0 & f_y & c_y \\ 0 & 0 & 1 \end{bmatrix}, \tag{2.7}
$$

where (X_w, Y_w, Z_w) is the 3D point in world coordinates, (X_c, Y_c, Z_c) is the same point in camera coordinates, (u, v) is the image in camera coordinates in pixels, $\mathbf{K}$ is a 5-DoF matrix with skew parameter k, and $[\mathbf{R} \ \mathbf{t}]$ is a 6-DoF matrix representing a rigid transformation composed of a 3-DoF 3D rotation and a 3-DoF 3D translation.

The matrix $\mathbf{K}[\mathbf{R} \ \mathbf{t}]$ is an 11-DoF matrix. Note that the DoF of this matrix becomes the same as that of a general projective transformation from a 3D space into a 2D space, which is $3 \times 4 - 1 = 11$.

The five free parameters of $\mathbf{K}$ represent the properties of camera projection. Thus, they are *intrinsic parameters*. On the contrary, a set of 6-DoF parameters representing the rigid transformation, $[\mathbf{R} \ \mathbf{t}]$, also represent the position and the pose of the camera in the world coordinate system. These parameters are *extrinsic parameters*. Note that there are many ways to represent a rigid transformation.

If intrinsic parameters of a camera are known, the 2D pixel position can be transformed into a 3D direction vector representing a ray in 3D space that is projected onto that pixel. This inverse calculation is often called back projection.

2.1.4   *Normalized Camera*

A camera in which $f_x = f_y = 1, k = c_x = c_y = 0$ is a normalized camera. Projection to a normalized camera can be described simply as

$$u_n = \frac{X_c}{Z_c}, v_n = \frac{Y_c}{Z_c} \qquad (2.8)$$

where (u_n, v_n) are the image coordinates on the image plane of a normalized camera.

As can be shown, (u_n, v_n) of the normalized camera can be transformed into image coordinates (u, v) of an arbitrary camera with projection matrix $\mathbf{K}$ by

$$\begin{bmatrix} u \\ v \\ 1 \end{bmatrix} = \mathbf{K} \begin{bmatrix} u_n \\ v_n \\ 1 \end{bmatrix}. \qquad (2.9)$$

Similarly, image coordinates (u, v) of arbitrary camera having a projection matrix, $\mathbf{K}$, can be transformed into (u_n, v_n) of a normalized camera by

$$\begin{bmatrix} u_n \\ v_n \\ 1 \end{bmatrix} = \mathbf{K}^{-1} \begin{bmatrix} u \\ v \\ 1 \end{bmatrix}. \qquad (2.10)$$

Thus, an image of a camera represented by a matrix $\mathbf{K}$ and that of a normalized camera can be transformed into each other using matrix $\mathbf{K}$.

With a normalized camera, the projection represented by the form of Eq. (2.7), can be rewritten as

$$s \begin{bmatrix} u_n \\ v_n \\ 1 \end{bmatrix} = \begin{bmatrix} X_c \\ Y_c \\ Z_c \end{bmatrix} = \begin{bmatrix} \mathbf{R} \ \mathbf{t} \end{bmatrix} \begin{bmatrix} X_w \\ Y_w \\ Z_w \\ 1 \end{bmatrix},$$

$$\begin{bmatrix} u \\ v \\ 1 \end{bmatrix} = \mathbf{K} \begin{bmatrix} u_n \\ v_n \\ 1 \end{bmatrix}. \qquad (2.11)$$

These forms show that a projection of a pinhole camera having projection matrix $\mathbf{K}$ can be interpreted as a composition of the projection of a normalized camera and the 2D homography represented by matrix $\mathbf{K}$.

2.1.5 Epipolar Geometry

Assume that there are two cameras, (cameras 1 and 2), and their relative positions are known. Thus, if a 3D point is represented as (X_1, Y_1, Z_1) in the camera coordinate system of camera 1, and as (X_2, Y_2, Z_2) in the camera coordinate system of camera 2, then the transformations between them is written as

$$
\begin{bmatrix} X_2 \\ Y_2 \\ Z_2 \\ 1 \end{bmatrix} = \begin{bmatrix} \mathbf{R} & \mathbf{t} \\ \mathbf{0}^\top & 1 \end{bmatrix} \begin{bmatrix} X_1 \\ Y_1 \\ Z_1 \\ 1 \end{bmatrix}. \tag{2.12}
$$

Also assuming that a 3D point is observed as a 2D point (u_1, v_1) for camera 1, and a 2D point (u_2, v_2) for camera 2, as shown in Fig. 2.3, the 3D point, the optical center of camera 1, and the optical center of camera 2 form a triangle, where the ray corresponding to (u_1, v_1), the ray corresponding to (u_2, v_2), and the baseline between cameras 1 and 2 are the edges.

Via back projection, the ray from camera 1 to point (X, Y, Z) has a direction vector

$$
\mathbf{R}\mathbf{K}_1^{-1} \begin{bmatrix} u_1 \\ v_1 \\ 1 \end{bmatrix} \tag{2.13}
$$

in the camera coordinate system of camera 2. The ray from camera 2 to point (X, Y, Z) has a direction vector

Fig. 2.3 Epipolar geometry

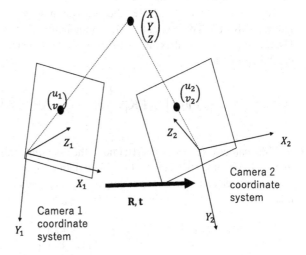

$$\mathbf{K}_2^{-1} \begin{bmatrix} u_2 \\ v_2 \\ 1 \end{bmatrix} \tag{2.14}$$

in the camera coordinate system of camera 2. The baseline corresponds to the translation vector, $\mathbf{t}$, in the camera coordinates of camera 2.

It is well-known that a scalar triple product of three points, $\mathbf{a}$, $\mathbf{b}$, and $\mathbf{c}$, defined as $\mathbf{a}^\top (\mathbf{b} \times \mathbf{c})$, becomes zero if $\mathbf{a}$, $\mathbf{b}$, and $\mathbf{c}$ are on a plane. Using this, we obtain

$$\begin{bmatrix} u_2 & v_2 & 1 \end{bmatrix} \mathbf{K}_2 \left(\mathbf{t} \times \mathbf{R}\mathbf{K}_1^{-1} \begin{bmatrix} u_1 \\ v_1 \\ 1 \end{bmatrix} \right) = 0. \tag{2.15}$$

The cross-product operation with $\mathbf{t}$ (the part "$\mathbf{t}\times$", which is a linear operation) can be written as a multiplication with a matrix defined from $\mathbf{t}$. By writing this matrix as $[\mathbf{t}]_\times$, the above equation leads to

$$\begin{bmatrix} u_2 & v_2 & 1 \end{bmatrix} \mathbf{K}_2[\mathbf{t}]_\times \mathbf{R}\mathbf{K}_1^{-1} \begin{bmatrix} u_1 \\ v_1 \\ 1 \end{bmatrix} = 0, \tag{2.16}$$

where $[\mathbf{t}]_\times$ is defined to be $\begin{bmatrix} 0 & -t_z & t_y \\ t_z & 0 & -t_x \\ -t_y & t_x & 0 \end{bmatrix}$ with $\mathbf{t} \equiv \begin{bmatrix} t_x & t_y & t_z \end{bmatrix}^\top$.

If coordinates (u_1, v_1) are fixed, Eq. (2.6), represents a line using variables (u_2, v_2) on the image plane of camera 2. This line is called an *epipolar line*. Similarly, if coordinates (u_2, v_2) are fixed, the same equation represents a line having variables (u_1, v_1) on the image plane of camera 1.

The matrix of $[\mathbf{t}]_\times \mathbf{R}$ is called the essential matrix and is often written as $\mathbf{E}$. The matrix of $\mathbf{K}_2[\mathbf{t}]_\times \mathbf{R}\mathbf{K}_1^{-1}$ is called the fundamental matrix and is often written as $\mathbf{F}$. Using the essential matrix, $\mathbf{E}$, or the fundamental matrix, $\mathbf{F}$, the epipolar constraint can be written as

$$\begin{bmatrix} u_2 & v_2 & 1 \end{bmatrix} \mathbf{K}_2\mathbf{E}\mathbf{K}_1^{-1} \begin{bmatrix} u_1 \\ v_1 \\ 1 \end{bmatrix} = \begin{bmatrix} u_2 & v_2 & 1 \end{bmatrix} \mathbf{F} \begin{bmatrix} u_1 \\ v_1 \\ 1 \end{bmatrix} = 0. \tag{2.17}$$

If the camera is a normalized camera, the essential matrix, $\mathbf{E}$, and the fundamental matrix, $\mathbf{F}$, becomes the same. Thus, if we define $(u_{n,1}, v_{n,1})$ as the 2D coordinates of the normalized camera corresponding to (u_1, v_1), and $(u_{n,2}, v_{n,2})$ is defined similarly, then

$$\begin{bmatrix} u_{n,2} & v_{n,2} & 1 \end{bmatrix} \mathbf{E} \begin{bmatrix} u_{n,1} \\ v_{n,1} \\ 1 \end{bmatrix} = 0. \tag{2.18}$$

2.1.6 Modeling Lens Distortions

A pinhole camera is the ideal model of a perspective projection camera. However, actual camera lenses have aberrations, and those aberrations cause image distortions. An important distortion model caused by aberration is radial distortion. The Brown–Conrady model of projection used in OpenCV [2, 3] is expressed by the following equation, modifying Eq. (2.11)

$$s \begin{bmatrix} v_n \\ u_n \\ 1 \end{bmatrix} = \begin{bmatrix} \mathbf{R} \ \mathbf{t} \end{bmatrix} \begin{bmatrix} X_w \\ Y_w \\ Z_w \\ 1 \end{bmatrix},$$

$$u' = u_n \frac{1 + k_1 r^2 + k_2 r^4 + k_3 r^6}{1 + k_4 r^2 + k_5 r^4 + k_6 r^6} + 2 p_1 v_n u_n + p_2 (r^2 + 2 u_n^2)$$

$$v' = v_n \frac{1 + k_1 r^2 + k_2 r^4 + k_3 r^6}{1 + k_4 r^2 + k_5 r^4 + k_6 r^6} + 2 p_2 v_n u_n + p_1 (r^2 + 2 v_n^2)$$

$$\text{where} \quad r^2 = u_n^2 + v_n^2,$$

$$\begin{bmatrix} u_d \\ v_d \\ 1 \end{bmatrix} = \mathbf{K} \begin{bmatrix} u' \\ v' \\ 1 \end{bmatrix}, \tag{2.19}$$

where the coordinate (u_d, v_d) represents an image point shifted by radial distortions. k_1, k_2, k_3, k_4, k_5, and k_6 are radial distortion coefficients, and p_1 and p_2 are tangential distortion coefficients. Note that an image of a pinhole camera having radial distortion (i.e., (u_d, v_d)) can be transformed into the image of a pinhole camera without radial distortion (i.e., (u, v)) by transforming (u, v) into the image of a normalized camera (u_n, v_n) and transforming (u_n, v_n) to (u, v) using the above forms. Using this mapping, a captured 2D image having a radial distortion can be pre-transformed into a corresponding 2D image without radial distortion.

2.2 Camera Calibration

In the previous section, mathematical forms of commonly used camera models were described. These models have many parameters. In many cases, these parameters must be determined. Camera calibration comprises the estimation of the intrinsic parameters of a certain camera or that of the relative positional information between cameras (i.e., extrinsic parameters). To retrieve 3D scene information from camera images, the intrinsic and extrinsic parameters of the cameras are needed. Thus, camera calibration is an imperative technique.

The most standard method of camera calibration uses calibration objects. Plane-shaped objects with checker patterns or grid-like black circles printed on the surface

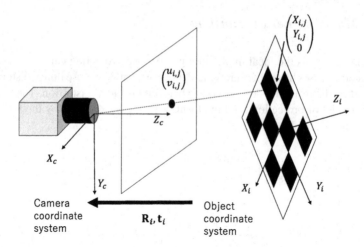

Fig. 2.4 Camera calibration

are commonly used. The OpenCV library and the Matlab Image Toolkit include function sets wherein feature points of such calibration objects can be detected and used.

Figure 2.4 shows an example of the camera calibration process. The plane-shaped calibration object is captured by a camera, and multiple images are captured while changing the angle of the calibration object. Ideally, tens of images must be captured at various angles to achieve calibrated parameter precision.

The positions of the feature points printed on the calibration object are a priori knowledge represented in an object coordinate system, which is a 3D coordinate system fixed to the calibration object. For example, when using a plane-shaped calibration object, the object coordinate system is often defined so that the object plane coincides with the xy plane of the object coordinate system wherein the directions of x and y-axes are aligned with the grid of the feature points.

Let the position of the j-th feature point of the i-th image be $(X_{i,j}, Y_{i,j}, 0)$ in the object coordinate system of the calibration object. The observed position of the feature point is at $(u_{i,j}, v_{i,j})$ on the i-th image. Thus,

$$
s \begin{bmatrix} u_{i,j} \\ v_{i,j} \\ 1 \end{bmatrix} = \mathbf{K} \begin{bmatrix} \mathbf{R}_i & \mathbf{t}_i \end{bmatrix} \begin{bmatrix} X_{i,j} \\ Y_{i,j} \\ 0 \\ 1 \end{bmatrix}.
\tag{2.20}
$$

The calibration process estimates $\mathbf{K}$, $\mathbf{R}_i$, and $\mathbf{t}_i$ so that Eq. (2.20), is most precisely fulfilled.

This calculation is often performed first with linear algebra [4], and then with nonlinear optimization using the linear solution as the initial solution. Users can use OpenCV or Matlab libraries for this calculation.

2.3 Projector Models

Geometrical modeling of pattern projection is an important part of triangulation-based active-light range sensing. In the following three subsections, three typical types of pattern projection modeling is explained.

2.3.1 Modeling Point-Laser Modules

Point-laser modules are used to triangulate a point. A key drawback of using a point-laser module for 3D measurement is that it provides information only for a single point. Thus, scanning systems are needed for measuring a 3D scene.

Land surveyors use point-laser modules, because position information of a small number of reference points is sufficient. A point-laser module projects a point onto a surface where the projected light travels a 3D path that includes the light-emitting point of the laser module and the point lit by it. By describing this 3D line using an arbitrary 3D coordinate system, we can model the point-laser module.

One example of describing a 3D line is as follows: Let the 3D line traverse through a point (X_l, Y_l, Z_l) and let the directional vector of the line be (X_d, Y_d, Z_d). Then, the points on the 3D line (X, Y, Z) can be described as

$$\begin{bmatrix} X \\ Y \\ Z \end{bmatrix} = \begin{bmatrix} X_l \\ Y_l \\ Z_l \end{bmatrix} + s \begin{bmatrix} X_d \\ Y_d \\ Z_d \end{bmatrix} \tag{2.21}$$

where s is an arbitrary real variable (Fig. 2.5).

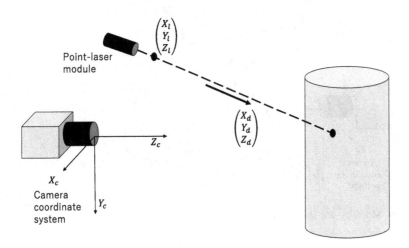

Fig. 2.5 Modeling point-laser modules

Sometimes, the line-laser module is described as a half line, noting that the laser light is beamed one way. In this case, s is assumed to be $s > 0$.

2.3.2   Modeling Line-Laser Modules

Line-laser modules are often used for practical 3D shape measurement. By combining a line-laser module, a camera, and a scanning system (e.g., motorized stage), we can construct a 3D scanning system based on the light-section method.

A line-laser module projects a straight line onto a 3D plane. In this book, we call it a laser plane. Thus, the projected light traverses 3D paths that include the light-emitting point of the laser module and the straight line projected onto the laser plane. By describing the laser plane using an arbitrary 3D coordinate system, we can model the line-laser module as shown in Fig. 2.6. Note that it is not necessary to use the coordinate system of the projecting device, because the geometry of the laser plane is simple. This is not the case of the video-pattern projector discussed in the next section.

One example of describing a laser plane is as follows: Let the laser plane go through a point (X_p, Y_p, Z_p) and let the normal vector of the plane be (X_n, Y_n, Z_n). Then, points (X, Y, Z) on the 3D line can be described by

$$\{[X \ Y \ Z] - [X_p \ Y_p \ Z_p]\} \begin{bmatrix} X_n \\ Y_n \\ Z_n \end{bmatrix} = 0. \tag{2.22}$$

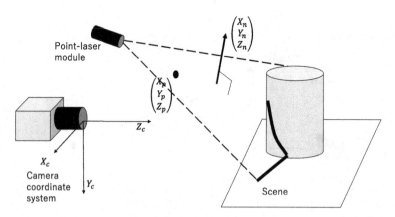

Fig. 2.6 Modeling line-laser modules

2.3.3   Projector Model of Pattern Projector

Projectors, especially those that project 2D patterns onto surfaces, can be generally modeled using the camera models described in Sect. 2.1. Regarding LCD video projectors, 2D images on the LCD plane are mapped onto a 3D surface. In this case, this includes the screen or other objects. The mapping between the 3D points on the surface and the 2D (LCD) plane can be represented the way as with the camera model shown in Fig. 2.7. Thus, the camera model formulation of Eq. (2.7), can also be used as a projector model. For example,

$$
s \begin{bmatrix} u_p \\ v_p \\ 1 \end{bmatrix} = \mathbf{K}_p \begin{bmatrix} X_p \\ Y_p \\ Z_p \end{bmatrix} = \mathbf{K}_p \begin{bmatrix} \mathbf{R}_p & \mathbf{t}_p \end{bmatrix} \begin{bmatrix} X_w \\ Y_w \\ Z_w \\ 1 \end{bmatrix} \tag{2.23}
$$

where (X_w, Y_w, Z_w) is the 3D point in world coordinates, (X_p, Y_p, Z_p) is the same point in projector coordinates, (u_p, v_p) is the location of the 2D point using the projected pattern in pixels, $\mathbf{K}_p$ is a 5-DoF matrix of intrinsic projector parameters, and $[\mathbf{R}_p \ \mathbf{t}_p]$ is a 6-DoF matrix representing extrinsic projector parameters.

A lens distortion model for cameras is described in Sect. 2.1.6, and can also be used for projectors.

For modeling real video projectors, the position of the principal point of the camera model should be treated carefully. In most cases of modeling normal cameras, the positions of the principal points can be put simply onto the center of the image plane. Some applications assume this. However, for off-the-shelf video projectors, the positions of the principal points are normally far from the centers of the image planes. For example, for a projector placed on a floor, the projected image should be

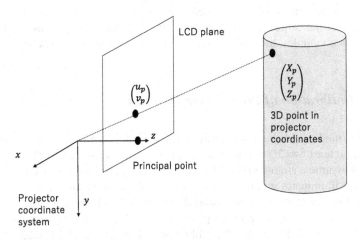

Fig. 2.7 Modeling video projectors

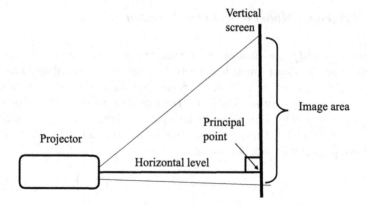

Fig. 2.8 Position of principal point of video projectors

higher than the optical axis, which is vertical to the screen, as shown in Fig. 2.8, so that the projected image is projected on the screen above the floor. Thus, the principal point should be near the bottom of the image plane. For projectors mounted on ceilings, they are set up upside-down. Thus, the principal points are near the tops of the image planes.

2.4 Projector Calibration

In this section, calibration of pattern projectors is described. One problem of calibrating pattern projectors versus cameras is that pattern projectors are not sensors. Consequently, projectors should be calibrated using other sensor devices such as cameras, whereas cameras can be calibrated by themselves. Thus, projector calibration often requires the calibration of cameras. In the following subsections, calibration of point-laser modules, line-laser modules, and video projectors are described.

2.4.1 Calibration of Point-Laser Modules

Point-laser modules can be described as 3D lines, as shown in Sect. 2.3.1. To calibrate a 3D line, at least two 3D points on the laser beam should be measured.

Let us assume a simple set up of a camera and a point-laser module, where the laser beam illuminates a point on a surface and the camera observes the point as shown in Fig. 2.9. The camera is assumed to have been calibrated in advance.

One simple way to calibrate the position of the laser beam is to use a camera calibration object. For example, the calibration object of a plane is placed such that the object is observed by the camera. The laser beam simultaneously projects a point

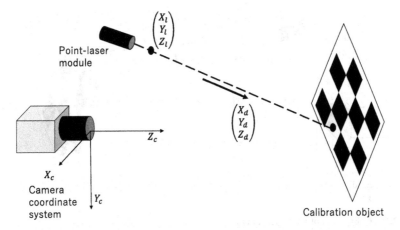

Fig. 2.9 Calibration of point-laser modules

onto the object. The point lit by the beam is observed by the camera as shown in Fig. 2.9. In this case, the position of the calibration object can be calculated in camera coordinates by using calibration libraries, such as OpenCV. Simultaneously, the 3D coordinates of the point lit by the laser beam can be calculated as the intersection of the plane, P, and the line, OL, of Fig. 2.9. OL goes through the origin of the camera coordinates, and its directional vector can be calculated by back-projecting the observed 2D point of L.

This process should be repeated at least twice, although many observations would lead to a more precise estimation of the laser beam. The measured 3D points are all on the 3D line of the beam. Thus, by estimating the line (e.g., via PCA), the laser beam can be calibrated.

2.4.2 Calibration of Line-Laser Modules

Line-laser modules can be described as a 3D plane, as shown in Sect. 2.3.2. To calibrate a 3D line, at least three 3D points on the 3D plane should be measured. Similar to point-laser modules, camera calibration objects can be used for estimating the 3D plane lit by a line-laser. In this case, the observed laser-lit points on the calibration object is a line. Like the case of the point-laser modules, the 3D points of the laser-lit points can be calculated as shown in Fig. 2.10. By collecting the 3D points on the 3D plane representing the line-laser, the 3D plane can be estimated (e.g., via PCA).

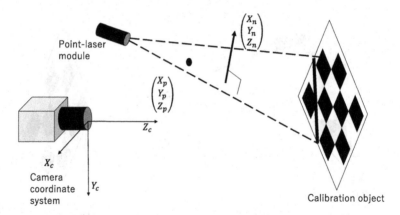

Fig. 2.10 Calibration of line-laser modules

2.4.3   Calibration of Video or 2D Pattern Projectors

Single-View Calibration Methods Using Specialized Calibration Objects

Considering cases where there is a projector-camera pair. As mentioned, the projector can be modeled using the same perspective projection formula as the camera. In this case, as with cameras, estimating internal and external parameters is important to many applications.

Unlike a camera, a projector is not a sensor that captures information from the real world. Thus, it cannot obtain information by itself. Therefore, sensors, such as cameras, are required for calibration.

Figure 2.11a shows an example projector calibration. In this example, the projector projects a calibration pattern onto a planar calibration object, and the image is captured by the camera. To locate the planar calibration object, calibration patterns (e.g., checkerboard or augmented-reality (AR) markers) are printed onto the object. The camera should capture both the printed and projected patterns.

Instead of projecting calibration markers, using structured light projection (e.g., Gray code patterns) enables us to obtain dense correspondence information between the projector pixels and the calibration object, as shown in Fig. 2.11b. Because denser correspondences are obtained, calibration accuracies are improved [5–7]. However, this approach requires fixing the calibration pattern while multiple Gray code patterns are projected and captured. This increases the calibration burden.

The typical approach requires a two-step algorithm, in which the intrinsic parameters of the camera are first estimated using a normal camera calibration method and a normal calibration object. Then, using the obtained camera parameters, for each of the captured image for the set up of Fig. 2.11a, the correspondences between the 3D point coordinates with the camera coordinates on the calibration object and the 2D pixels of the projector can be obtained [8–12]. From the correspondence pairs of 3D

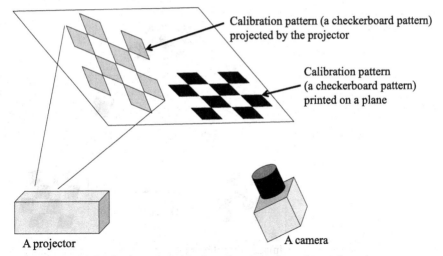

(a) Projection of checkerboard pattern from the projector.

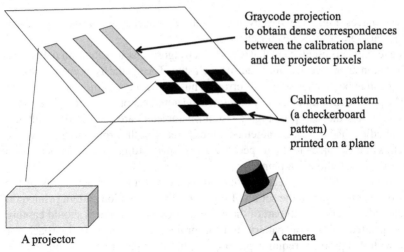

(b) Projection of Gray code patterns for dense correspondences.

Fig. 2.11 Calibration of a video projector

and 2D coordinates, projector calibration can be achieved. Freely available software packages for this approach can be found [12, 13].

Some researchers use 2D correspondences for synthesizing images that would have been captured if the projector had been a camera (i.e., virtual projector images) [5, 6, 14, 15]. Then, they applied well-known camera calibration techniques to synthesized images, because these images could be regarded the same way

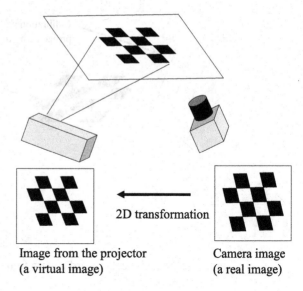

Fig. 2.12 Synthesis of projector images

2D transformation

Image from the projector   Camera image
(a virtual image)   (a real image)

as camera images (Fig. 2.12). An advantage of this approach is that we can use widely available calibration softwares for cameras.

Some researchers pointed out that the two-step algorithms are suboptimal, because the calibration errors are accumulated for each step. The synthesis of virtual projector images may be another source of errors. Thus, camera and projector calibration should be processed simultaneously or refined via nonlinear optimization [7, 15–17]. For example, for the set up of Fig. 2.11a, all parameters are estimated simultaneously, including the intrinsic parameters of the camera and the projector, the extrinsic parameters of the projector with respect to the camera coordinates, and the positional parameters of the calibration plane.

For accurate calibration, feature points should cover all the image areas of the camera or the projector. In the set up of Fig. 2.13, the calibration pattern printed on the plane only covers a small part of the camera, because the camera should capture both the printed calibration pattern and the projected pattern. One approach for dealing with this problem requires printing a calibration pattern in red and another in blue, overlapping them. From the captured images, overlapped patterns can be separated by extracting red and blue color channels.

Single-View Calibration Methods Using Ordinary Objects or Natural Scenes

Providing a specialized calibration object is sometimes difficult. Some researchers have proposed projector calibration methods in which specialized calibration patterns are unnecessary. Some use a simple white plane to calibrate multiple poses of the projector [18, 19]. In this set up, the user moves the projector while capturing the projection images.

One problem of using a white plane for a calibration object is that the captured images are scale-free. In other words, we cannot distinguish between an image where

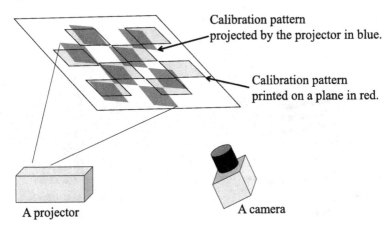

Calibration pattern
projected by the projector in blue.

Calibration pattern
printed on a plane in red.

A projector   A camera

Fig. 2.13 Overlapped calibration patterns. One pattern is printed on the plane and another different pattern is projected onto the plane

a plane and a projector with certain positions, and another where the positions of the plane and the projector are scaled with respect to the camera position by, for example, 10 times.

One solution for the scale problem is using a calibration object that is not scale-free. For example, a method has been proposed in which the intrinsic and extrinsic parameters of a projector can be obtained by projecting a pattern onto a sphere and observing it with a camera [20]. Because spheres are not scale-free, we can determine the projector's position using a real scale.

Using structured projection, dense correspondences between the projector and the camera can be obtained. Considering that these correspondences are the same as the correspondences between a stereo-camera pair, stereo calibration techniques can be used for this system without the need for special calibration objects. Calibrating the system directly from the measured scene is called self- or auto-calibration. Several self-calibration methods have been proposed [16, 21–24]. The calibration software developed by Yamazaki et al. [16], is publicly available.

Furukawa et al. [21] proposed a shape scanning system based on this approach. In this approach, a user can set up a camera and a projector freely in front of the scene, and scan the scene using Gray code projection. Then, the extrinsic parameters are automatically calculated and the shape of the scene is reconstructed. An example is shown in Fig. 2.14

With those scanning systems using self-calibration, calibration accuracy may be affected by the geometrical structure of the scene. For example, the extreme case in which the scene consists of only a single plane would cause a degenerate condition, causing unreliable calibration results.

Note that self-calibration can calibrate the projector-camera system only up to scale. Thus, the scale of the reconstructed scene remains ambiguous. Thus, if a shape measurement with a real scale is needed, other constraints, such as known-sized

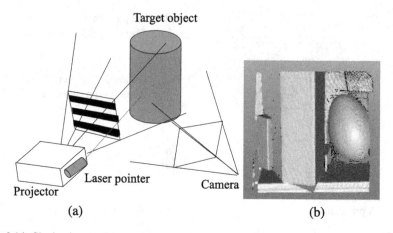

Target object

Laser pointer

Projector   Camera

(a)   (b)

Fig. 2.14 Single-view projector-camera system using self-calibration [21]. **a** System setting.
b Scanning result

shape, or other equipment, such as laser projector mounted on the projector-camera
system, should be added [21], as shown in Fig. 2.14.

Multi-view Calibration Methods

A single-view projector-camera system can only capture one side of an object, but
not the entire object or the scene. To address this problem, researchers have proposed
multi-view projector-camera systems [21, 23–28].

For these, acquiring correspondences between different viewpoints is a serious
problem. In a passive multi-view stereo system, correspondences can be obtained by
finding similar key points from images from different viewpoints. Similar key points
can be detected by comparing feature vectors of the key points using algorithms,
such as scale-invariant feature transform (SIFT) or oriented FAST and rotated BRIEF
(ORB) features [29, 30].

In most active multi-view systems, it is assumed that the correspondences are
obtained only from structured light projections (e.g., Gray codes), because the existence of textures or geometrical features cannot be assumed. In these cases, only
correspondences between a camera and a nearby projector can be obtained. For
example, Fig. 2.15, shows an example of a multi-view projector-camera system. In
this setup, projectors (P1, P2, ...) and cameras (C1, C2, ...) are alternatively placed
around the target scene. By using structured light projection, dense correspondences
between the adjacent devices (e.g., P1 and C1, C1 and P2, P2 and C2, C2 and P3)
are obtained.

In such cases, correspondences between different views are obtained via chaining. Figure 2.16 illustrates the concept of chaining correspondences. In this situation,
dense correspondences between C1 and P2 are obtained using a structured light projection. Similarly, correspondences between P2 and C2 are already known. Thus,
from these correspondences, by using a pixel on P2 as a pivot, a correspondence

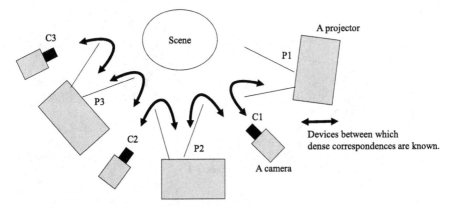

Fig. 2.15 Multi-view projector-camera system

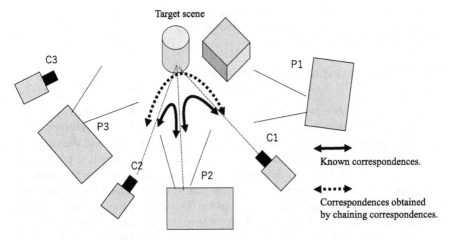

Fig. 2.16 Obtaining correspondences for a multi-view projector-camera system by chaining correspondences

between C1 and C2 can be obtained. Similarly, correspondences between P2 and P3 can be obtained using a pixel of P1 as a pivot point. Using these techniques, the information of correspondence point pairs between different devices can be increased. When sufficient numbers of correspondences are obtained, sparse bundle-adjustment techniques [31] can be applied in a way similar to passive multi-view stereo. The concept of chaining can be seen in Furukawa et al. [21], as a pivot scanning, which was proposed for scanning an entire object, as shown in Fig. 2.17.

In cases of multi-view passive stereo, the greater the number of viewpoints, the better the reconstruction accuracy, because geometrical constraints increase with the number of viewpoints. Similarly, multi-view projector-camera systems tend to be more accurate than single-viewpoint projector-camera systems.

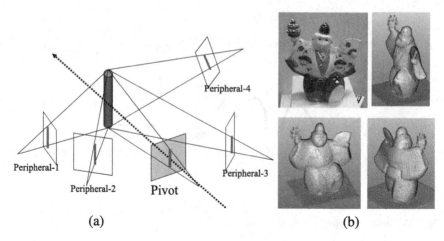

Fig. 2.17 Merging multiple shapes scanned using active stereo [21]. **a** Pivot scanning. **b** Entire shape reconstruction of an object using pivot scanning

2.5 Point Correspondence Problem

In a reconstruction using structured light, 3D reconstruction is performed by capturing single or multiple patterns projected onto a scene. To realize the 3D reconstruction, correspondence between the point on the projected pattern and the captured image is required. A typical system configuration and the corresponding-point problem are shown in Fig. 2.18a. As described, because the optical system of the projector is the same as that of the camera, the measurement system is equivalent to replacing one camera in the stereo system with a projector. Therefore, epipolar constraints can be used to solve the point correspondence problem, and the search range is limited to 1D along the epipolar line. If this property is used, it is possible to use a 1D pattern (e.g., stripes).

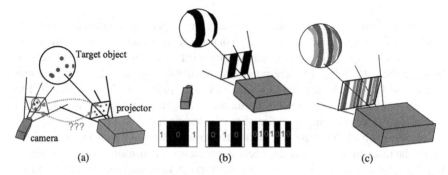

Fig. 2.18 Point correspondence problem of a projector-camera system. **a** System configuration and correspondence problem. **b** Temporal encoding method. **c** Spatial encoding method

To obtain the correspondence, one approach is to embed the position information using patterns that change temporally or spatially [32]. Temporal encoding projects multiple patterns, as shown in Fig. 2.18 (middle), and embeds location information into the temporal change. However, as shown in Fig. 2.18b, the spatial encoding method embeds position information in a local spatial region. A method using only temporal change is easy to implement, high in precision, high in density, and stable. Thus, it is widely used for real applications. However, because this method requires multiple images, it is not suitable for fast reconstruction. Another approach is to encode the information for identifying the position of each pixel using a spatial pattern in the image. The 3D reconstruction of a scene is realized by capturing only one image. This approach is suitable for the reconstruction of moving objects for which decoding temporal patterns from multiple images is difficult when searching for point correspondence. However, because spatial encoding requires an area larger than one pixel to encode the position information, the appearance of the pattern can change if the target object has a complicated shape or color. Thus, it is difficult to find a correspondence in such cases.

In the following sections, these two typical encoding methods are introduced, and their extended methods are described in Sect. 6.2.

2.5.1 Temporal Encoding

Temporal encoding of a pattern is accomplished by embedding the position information using multiple patterns. Because the method of utilizing temporal changes of an image is easy to implement, it has been used for a long time [33–37]. In the following subsections, the binary code pattern projection and phase-shifting methods are explained as basic temporal encoding methods.

Binary Code Pattern Projection

The oft-used temporal encoding method projects a black and white binary stripe pattern, as shown in Fig. 2.18b. Let the x-coordinate of each pixel in the pattern to be projected be represented by an N-bit code, $\mathbf{c} = (c_0, c_1, \ldots, c_{N-1})$. Simultaneously, the pattern is divided into N images representing each bit, and each pixel of the i-th image is the i-th bit from the most significant bit of $\mathbf{c}$. It is determined by the value, c_i, of black, if $c_i = 0$, or white, if $c_i = 1$. By projecting the N patterns in a time series to the scene and determining if a pixel is black or white, a binary sequence of N bits for each pixel, $\mathbf{d} = (d_0, d_1, \ldots, d_{N-1})$. If there is no error in observation, $\mathbf{d} = \mathbf{c}$, and the projector coordinates, x, corresponding to $\mathbf{c}$, can be uniquely determined. Furthermore, the correspondence between the pixel and the projector coordinates is obtained.

A simple way to create $\mathbf{c}$ from coordinate values is to use the binary code of x. However, this method is vulnerable to errors. For example, if c_0 is incorrect when $N = 4$, x deviates 4 from the correct value. The Gray codes are often used as binary sequences to avoid this problem. Let $\mathbf{b} = (b_0, b_1, \ldots, b_{N-1})$ and $\mathbf{g} = (g_0, g_1, \ldots, g_{N-1})$: the binary and Gray code representation. The conversion from **b** to **g** is expressed by

$$g_0 = b_0, \quad g_{i+1} = b_i \oplus b_{i+1}, \tag{2.24}$$

Inversely, the conversion from **g** to **b** is

$$b_0 = g_0, \quad b_{i+1} = b_i \oplus g_{i+1} \tag{2.25}$$

Thus, the binary code 0101 corresponds to the gray code 0111.

The Gray code has a feature for which code $c(x)$ at coordinate x and the one, $c(x + 1)$, of the adjacent coordinate, $x + 1$, differ by only one bit. This feature provides two advantages. One is that the boundary of the pattern, where the error of discriminating black and white is likely to occur, exists at only one image between $c(x)$ and $c(x + 1)$. Additionally, the number of boundaries in the whole pattern is minimized. The other feature is that, even if a discriminating error occurs at the pattern boundary, the error of the calculated coordinate, x, is suppressed to only 1, because the incorrect c' value is the pattern next to the correct pattern, **c**. Figure 2.19 shows the pattern via binary code and Gray code for $N = 4$. The leftmost image represents the least significant bit, and the rightmost image represents the most significant. By observing four images, different patterns of 2 bits are obtained in the horizontal axis direction, and coordinates 0–15 are assigned.

The Gray code has the feature that the code $c(x)$ at coordinate x and the one $c(x + 1)$ of adjacent coordinate $x + 1$ have difference by only one bit. This feature has two advantages. One is that the boundary of the pattern, where the error of discriminating black and white is likely to occur, exists at only one image between $c(x)$ and $c(x + 1)$, and the number of boundaries in the whole pattern is minimized. The other is that even if a discriminating error occurs at the pattern boundary, the error of the calculated coordinate x is suppressed to only 1 because the incorrect c' is the pattern next to the correct pattern **c**. Figure 2.19 shows the pattern by binary code and gray code for $N = 4$. The leftmost image represents the least significant bit, and the rightmost image represents the most significant bit. By observing four images, different patterns of 4 bits are obtained in the horizontal axis direction, and coordinates from 0 to 15 are assigned.

Using Gray codes, the correspondence between integer coordinates can be obtained between camera and projector images. If the shape is calculated using the integer coordinates, quantization errors will occur, and the shape will be reconstructed in a step-like fashion. To obtain an accurate shape, it is necessary to find correspondences at sub-pixel accuracy.

Sato and Inokuchi [38] proposed a method of assigning the x-coordinate value of a projector to the boundary of the projector pixel and detecting the boundary position in the camera image with sub-pixel accuracy. Therefore, instead of obtaining the corresponding projector coordinates for each pixel of the camera, the camera coordinates corresponding to each projector's x coordinate are calculated. To stably detect the boundary position, negative patterns are projected in which black and white, inverted on each projection pattern. Because the brightness value is inverted

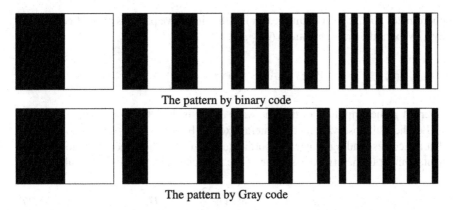

The pattern by binary code

The pattern by Gray code

Fig. 2.19 Example of binary and Gray code patterns for $N = 4$

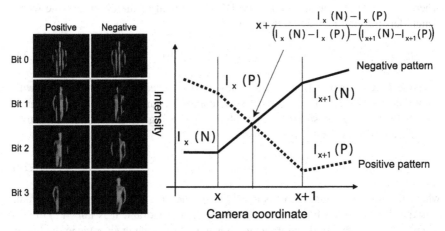

Fig. 2.20 Camera coordinates corresponding to the boundaries of projector pixels are calculated with sub-pixel accuracy by projecting positive and negative patterns

at the pixel boundary, the position where the brightness value of the positive/negative patterns intersects in the graph is calculated as the boundary, as shown in Fig. 2.20.

Phase Shifting

Phase shifting is often used with temporal encoding methods. In this method, periodic patterns (e.g., sinusoidal) are projected onto the observation target multiple times with phase shifting. By observing the change in the brightness value associated with the phase, it is possible to calculate the phase of the period for each pixel. Because the phase is determined using the brightness value, the correspondence between camera and projector can be obtained with sub-pixel accuracy.

First, the method used to calculate the phase from the brightness values using $N (\geq 3)$ patterns is described. If the spatial frequency of the pattern with respect to the normalized projector coordinates is f, the phase of the sine wave at projector coordinates x is $2\pi f x$. Let this be $\phi(x)$. If the sine wave patterns are shifted N times by $1/N$ periods, the n-th pattern $(0 \leq n \leq N - 1)$ is

$$P_n(x) = A \cos\left(\phi(x) + \frac{2\pi n}{N}\right) + B,   \tag{2.26}$$

where A is the amplitude of the sine wave and B is the DC component.

Let the n-th observation at a camera pixel, p, be I_n $(0 \leq n \leq N - 1)$. Assuming that the corresponding projector coordinates are x, and the direct reflection component is linear to the brightness value of the incident light, the observed value is

$$I_n = C \cos\left(\phi(x) + \frac{2\pi n}{N}\right) + D,   \tag{2.27}$$

where C and D are the amplitude and DC component of the observed sine wave, which depends on the pixel p. Because

$$I_n = \cos(2\pi n/N)(C \cos(\phi(x))) - \sin(2\pi n/N)(C \sin(\phi(x))) + D,   \tag{2.28}$$

by using $E \equiv C \cos(\phi(x))$ and $F \equiv C \sin(\phi(x))$, N linear equations are given with the unknowns, D, E, and F. If $N = 3$, the equations can be solved directly, and if $N > 3$, the least-square solution can be obtained by using the pseudo-inverse matrix. The phase, $\phi(x)$, at pixel p is obtained by

$$\phi(x) = \mathrm{atan2}(E, F),   \tag{2.29}$$

where the function, $\mathrm{atan2}(x, y)$ is $\tan^{-1}(y/x)$ when (x, y) is in the first quadrant, it is $\tan^{-1}(y/x) + \pi$ when (x, y) is in the second quadrant, it is $\tan^{-1}(y/x) - \pi$ when (x, y) is in the third quadrant, and it is $\tan^{-1}(y/x)$ when (x, y) is in the fourth quadrant. An example for $N = 3$ is shown in Fig. 2.21.

$\phi(x)$ represents periodic relative coordinates with respect to the projector coordinates. Although the correspondence within each cycle can be determined uniquely, the ambiguity of the cycle remains. Therefore, it is necessary to remove ambiguity using a phase-unwrapping method to uniquely associate each camera pixel with projector coordinates. Various methods of phase unwrapping have been proposed, such as using the integration order [37], combining with binary code [39], and using multiple-wavelength periodic patterns [40]. One issue with these methods is that the number of images is increased. Recently, however, a method of realizing phase unwrapping by using the constraint condition between only phase images in both vertical and horizontal directions has been proposed [41].

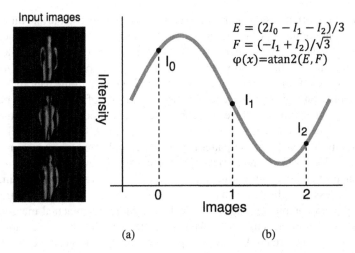

(a)   (b)

Fig. 2.21 Phase-shift patterns. **a** Capture the sine wave patterns three times with phase shifting $\frac{2}{3}\pi$. **b** The brightness change of each pixel is sine wave shifted by the phase, $\phi(x)$, which can be determined by sampling three times

2.5.2 Spatial Encoding

Methods based on only spatial encoding [42–45] can measure 3D shapes using only a single image. This approach is suitable for capturing objects in fast motion, because it requires only a single image. However, the pattern of spatial encoding is generally complicated, owing to the need to uniquely encode position information using a local pattern around each pixel. Because single-position information requires multiple pixels, a problem arises wherein the resolution of the 3D reconstruction is lowered. Additionally, when the code pattern becomes complicated, the reflectance of the object can disturb the color information, or the shape of the object can cause distortion or discontinuity of the pattern. Thus, the extraction of encoded information becomes difficult and unstable. Consequently, when changes in object color or shape unevenness is large, the extraction of information is likely to fail. Thus, the accuracy of the shape is lowered. Methods for solving these problems and realizing the measurement of moving objects via spatial encoding are described in detail in Sect. 6.2.2, as one-shot methods.

The temporal encoding method introduced in Sect. 2.5.1, has been studied and commercialized thus far, because it can achieve relatively high accuracy with a simple configuration. However, the spatial encoding method that uses special patterns is easily affected by the shape and texture of the target object. Thus, it is difficult to improve accuracy and density. However, recently, shape measurement using only a single-frame image has attracted attention, because it can measure moving objects. The measurement method using a single-frame image is called one-shot scan.

The most important task of active measurement is obtaining stable and unambiguous corresponding points with sufficient density. In the 3D shape measurement system that uses structured light, there is a 1D ambiguity on the epipolar line when finding corresponding points. To solve this problem using the spatial encoding method, correspondence information is embedded into the 2D layout of the pattern.

For this reason, it is necessary to increase the density of the pattern to perform high-density measurement. However, in the case of a high-density pattern, the same pattern can appear repeatedly, which increases ambiguity. Generally, to eliminate the ambiguity of matching, the pattern becomes complicated when increasing the number of colors or using the intricate pattern shape. However, if the effects of the surface shape and the texture of the target object is considered, a smaller number of colors and a less-complicated pattern is better for accuracy. Thus, there is a trade-off between pattern complexity and accuracy. Furthermore, in an actual measurement scene, pattern distortion is inevitable when projected on a 3D object and captured as a 2D image. An important objective of spatial encoding is to generate an efficient pattern that simultaneously satisfies the contradicting demands of high density and low ambiguity under the severe conditions described above.

The methods that have been proposed as such spatial encoding methods so far can be roughly classified into the following three methods according to the methods of finding corresponding points. The outline of each method is shown in Fig. 2.22. The details of each method are introduced below.

Basic Idea

One-shot methods and the temporal encoding methods use the same setting that one of the cameras of a stereo set up is replaced with a projector. Thus, the problem of finding corresponding points is an essential issue. However, the given one-shot methods have only one input image, making the problem more severe. The benefits of shape reconstruction from a single image are great, and the following three solutions have been proposed.

1. **Line-based methods**

   This approach projects a line pattern and determines the correspondence via the intersection between projected and epipolar lines. Because the correspondence between the detected line and one on the projected pattern is separately required, it is necessary to embed the information in the line. An example of this approach is shown in Fig. 2.22a.

2. **Area-based methods**

   This approach determines the correspondence using a pattern in the local window. Therefore, it is necessary to embed unique information into each local area of the projected pattern. Originally, this method could obtain the correspondence without requiring the epipolar constraint. However, in practice, the corresponding point is often searched while moving the window on the epipolar line. In that case, the uniqueness of the local area pattern only needs to be realized on each epipolar line. Thus, encoding becomes easy. An example of this method is shown in Fig. 2.22b.

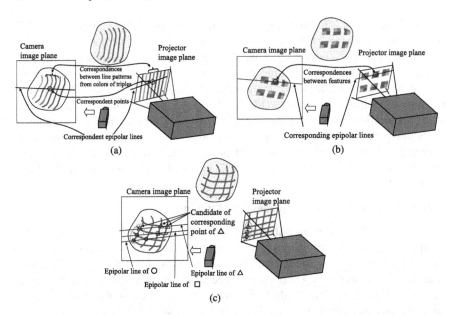

Fig. 2.22 One-shot encoding methods. **a** Line-based method. In the figure, the correspondence between lines is determined by the color of three consecutive lines. **b** Area-based method. Correspondence is determined by matching 2D features consisting of four colors. The feature matching is performed along epipolar lines. **c** Grid-based method. There are multiple candidates for corresponding points of ∘, □, and △ along the epipolar line. The green dotted line indicates the position of the correct corresponding point. (Color figure online)

3. **Grid-based methods**

   This approach determines the correspondence by using the connection between line intersections. This approach makes the pattern very simple, because no information is embedded in the line or the area. This feature is suitable for measuring moving objects. An example is shown in Fig. 2.22c.

Relationship with Epipolar Geometry

In all methods, the epipolar constraints are important for solving the correspondence problem. To solve the 2D corresponding-point problem using line-based methods, a 1D search is performed using epipolar constraints, and the remaining dimension is used for encoding the pattern. In other words, the epipolar constraint is why projection patterns (e.g., stripe patterns with the same color in the vertical direction) have been used. There are two cases for encoding methods with 1D. One is performed along a line, and the other is performed along an epipolar line. Because the projected pattern is deformed on the object's surface, it is necessary to solve by using elastic matching (e.g., dynamic programming). Thus, there is no need to consider expansion and contraction for temporal encoding.

In area-based methods, because 2D matching is possible, the 2D correspondence problem can be directly solved without using epipolar constraints. The corresponding points obtained in this fashion can be used for self-calibration [21]. Because the 3D measurement has been calibrated, the corresponding-point problem is solved by performing area (window) matching along the epipolar line using epipolar constraints.

Epipolar constraints are also important for solving the corresponding-point problem of grid-based methods. For the intersection of lines, there is a 1D-constraint equation for epipolar constraint. When the density of intersections is high, there are many intersections on the epipolar line. However, the intersections must simultaneously satisfy both constraints given by their connectivity and the epipolar constraint. Therefore, when the number of connected intersections increases, many epipolar constraints can be used simultaneously. This makes it possible to solve the corresponding-point problem.

Pattern Complexity and Uniqueness of Matching

A common problem with one-shot methods is the uniqueness of matching to the complexity of the pattern. There is one degree of freedom along the epipolar line in triangulation-based methods, and the information is encoded into the pattern to find the correspondence. Although it is necessary to increase the density of the pattern in order to perform high-density measurement, the possibility of the same pattern appearing repeatedly increases, which increases ambiguity in the case of a high-density pattern. Generally, a pattern is complicated by increasing the number of colors or using the shape of the pattern to eliminate the ambiguity of matching. However, considering the accuracy of image processing and the influence of the surface shape and texture of the target object, it is desirable to use a smaller number of colors or pattern shapes. Thus, there is a trade-off between pattern complexity, resolution, and accuracy, which is an important issue in designing one-shot methods.

Furthermore, although one-shot methods require spatial encoding of information, the target to be projected is a 3D object. Thus, it is captured in 2D by the camera. It is inevitable that the pattern becomes distorted. Solutions are generally unstable with simple matching because of the distortion. Various methods have, therefore, been proposed to tackle this issue, such as dynamic programming for matching or distortion-invariant patterns.

References

1. Bradski G (2000) The openCV library. Dr. Dobb's J Softw Tools
2. Brown DC (1966) Decentering distortion of lenses. Photogramm Eng Remote Sens
3. Conrady AE (1919) Decentred lens-systems. Monthly Not R Astron Soc 79(5):384–390
4. Zhang Z (2000) A flexible new technique for camera calibration. IEEE Trans Pattern Anal Mach Intell (PAMI) 22
5. Zhang S, Huang PS (2006) Novel method for structured light system calibration. Opt Eng 45(8):083601
6. Li Z, Shi Y, Wang C, Wang Y (2008) Accurate calibration method for a structured light system. Opt Eng 47(5):053604

7. Moreno D, Taubin G (2012) Simple, accurate, and robust projector-camera calibration. In: 2012 second international conference on 3D imaging, modeling, processing, visualization & transmission. IEEE, pp 464–471

8. Sadlo F, Weyrich T, Peikert R, Gross M (2005) A practical structured light acquisition system for point-based geometry and texture. In: Proceedings Eurographics/IEEE VGTC symposium point-based graphics. IEEE, pp 89–145

9. Liao J, Cai L (2008) A calibration method for uncoupling projector and camera of a structured light system. In: 2008 IEEE/ASME international conference on advanced intelligent mechatronics. IEEE, pp 770–774

10. Yamauchi K, Saito H, Sato Y (2008) Calibration of a structured light system by observing planar object from unknown viewpoints. In: Proceedings of the international conference on pattern recognition (ICPR). IEEE, pp 1–4

11. Wei G, Liang W, Zhan-Yi H (2008) Flexible calibration of a portable structured light system through surface plane. Acta Autom Sin 34(11):1358–1362

12. Audet S, Okutomi M (2009) A user-friendly method to geometrically calibrate projector-camera systems. In: Proceedings of the IEEE conference on computer vision and pattern recognition (CVPR) workshops. IEEE, pp 47–54

13. Falcao G, Hurtos N, Massich J, Fofi D (2009) Projector-camera calibration toolbox. Erasumus Mundus masters in vision and robotics

14. Anwar H, Din I, Park K (2012) Projector calibration for 3d scanning using virtual target images. Int J Precis Eng Manuf 13(1):125–131

15. Zhang X, Zhang Z, Cheng W (2015) Iterative projector calibration using multi-frequency phase-shifting method. In: 2015 IEEE 7th international conference on cybernetics and intelligent systems (CIS) and IEEE conference on robotics, automation and mechatronics (RAM). IEEE, pp 1–6

16. Yamazaki S, Mochimaru M, Kanade T (2011) Simultaneous self-calibration of a projector and a camera using structured light. In: Proceedings of the IEEE conference on computer vision and pattern recognition (CVPR) workshops. IEEE, pp 60–67

17. Shahpaski M, Ricardo Sapaico L, Chevassus G, Susstrunk S (2017) Simultaneous geometric and radiometric calibration of a projector-camera pair. In: Proceedings of the IEEE conference on computer vision and pattern recognition (CVPR), pp 4885–4893

18. Okatani T, Deguchi K (2005) Autocalibration of a projector-camera system. IEEE Trans Pattern Anal Mach Intell (PAMI) 27(12):1845–1855

19. Draréni J, Roy S, Sturm P (2009) Geometric video projector auto-calibration. In: Proceedings of the IEEE conference on computer vision and pattern recognition (CVPR) workshops. IEEE, pp 39–46

20. Furukawa R, Masutani R, Miyazaki D, Baba M, Hiura S, Visentini-Scarzanella M, Morinaga H, Kawasaki H, Sagawa R (2015) 2-dof auto-calibration for a 3d endoscope system based on active stereo. In: 2015 37th annual international conference of the IEEE engineering in medicine and biology society (EMBC). IEEE, pp 7937–7941

21. Furukawa R, Kawasaki H (2005) Uncalibrated multiple image stereo system with arbitrarily movable camera and projector for wide range scanning. In: Fifth international conference on 3-D digital imaging and modeling (3DIM'05). IEEE, pp 302–309

22. Albarelli A, Cosmo L, Bergamasco F, Torsello A (2014) High-coverage 3d scanning through online structured light calibration. In: Proceedings of the international conference on pattern recognition (ICPR). IEEE, pp 4080–4085

23. Fleischmann O, Koch R (2016) Fast projector-camera calibration for interactive projection mapping. In: Proceedings of the international conference on pattern recognition (ICPR). IEEE, pp 3798–3803

24. Li C, Monno Y, Hidaka H, Okutomi M (2019) Pro-cam ssfm: Projector-camera system for structure and spectral reflectance from motion. In: Proceedings of the international conference on computer vision (ICCV), pp 2414–2423

25. Furuakwa R, Inose K, Kawasaki H (2009) Multi-view reconstruction for projector camera systems based on bundle adjustment. In: Proceedings of the IEEE conference on computer vision and pattern recognition (CVPR) workshops. IEEE, pp 69–76

26. Furukawa R, Sagawa R, Kawasaki H, Sakashita K, Yagi Y, Asada N (2010) One-shot entire shape acquisition method using multiple projectors and cameras. In: Pacific-rim symposium on image and video technology (PSVIT). IEEE, pp 107–114

27. Kasuya N, Sagawa R, Furukawa R, Kawasaki H (2013) One-shot entire shape scanning by utilizing multiple projector-camera constraints of grid patterns. In: Proceedings of the international conference on computer vision (ICCV) workshops, pp 299–306

28. Willi S, Grundhöfer A (2017) Robust geometric self-calibration of generic multi-projector camera systems. In: 2017 IEEE international symposium on mixed and augmented reality (ISMAR). IEEE, pp 42–51

29. Lowe DG (2004) Distinctive image features from scale-invariant keypoints. Int J Comput Vis (IJCV) 60(2):91–110

30. Rublee E, Rabaud V, Konolige K, Bradski G (2011) Orb: an efficient alternative to sift or surf. In: Proceedings of the international conference on computer vision (ICCV). IEEE, pp 2564–2571

31. Lourakis MI, Argyros AA (2009) Sba: a software package for generic sparse bundle adjustment. ACM Trans Math Softw (TOMS) 36(1):1–30

32. Batlle J, Mouaddib E, Salvi J (1998) Recent progress in coded structured light as a technique to solve the correspondence problem: a survey. Pattern Recogn 31(7):963–982

33. Inokuchi S (1984) Range imaging system for 3-d object recognition. In: Proceedings of the international conference on pattern recognition (ICPR), pp 806–808

34. Caspi D, Kiryati N, Shamir J (1998) Range imaging with adaptive color structured light. IEEE Trans Pattern Anal Mach Intell (PAMI) 20(5):470–480

35. Boyer KL, Kak AC (1987) Color-encoded structured light for rapid active ranging. IEEE Trans Pattern Anal Mach Intell (PAMI) 1:14–28

36. Abdul-Rahman HS, Gdeisat MA, Burton DR, Lalor MJ, Lilley F, Moore CJ (2007) Fast and robust three-dimensional best path phase unwrapping algorithm. Appl Opt 46(26):6623–6635

37. Ghiglia DC, Pritt MD (1998) Two-dimensional phase unwrapping: theory, algorithms, and software, vol 4. Wiley, New York

38. Sato K (1987) Range imaging system utilizing nematic liquid crystal mask. In: Proceedings of the international conference on computer vision (ICCV), pp 657–661

39. Gühring J (2000) Dense 3d surface acquisition by structured light using off-the-shelf components. In: Videometrics and optical methods for 3D shape measurement, vol 4309. International Society for Optics and Photonics, pp 220–231

40. Zhao H, Chen W, Tan Y (1994) Phase-unwrapping algorithm for the measurement of three-dimensional object shapes. Appl Opt 33(20):4497–4500

41. Sagawa R, Kawasaki H, Kiyota S, Furukawa R (2011) Dense one-shot 3d reconstruction by detecting continuous regions with parallel line projection. In: Proceedings of the international conference on computer vision (ICCV). IEEE, pp 1911–1918

42. Je C, Lee SW, Park RH (2004) High-contrast color-stripe pattern for rapid structured-light range imaging. In: Proceedings of the European conference on computer vision (ECCV). Springer, pp 95–107

43. Tajima J, Iwakawa M (1990) 3-d data acquisition by rainbow range finder. In: Proceedings of the international conference on pattern recognition (ICPR), vol 1. IEEE, pp 309–313

44. Pan J, Huang PS, Chiang FP (2005) Color-coded binary fringe projection technique for 3-d shape measurement. Opt Eng 44(2):023606

45. Salvi J, Batlle J, Mouaddib E (1998) A robust-coded pattern projection for dynamic 3d scene measurement. Pattern Recogn Lett 19(11):1055–1065

Chapter 3
Sensor

To retrieve light and reflectance information from the real world, a wide variety of sensors have been developed for specific purposes. Among them, the RGB camera, which is the most similar to the human eye, is the most important. The analog camera was invented more than 100 years ago, and the film was recently replaced by digital imaging devices, (e.g., CCD or complementary metal oxide semiconductor (CMOS)). Another important sensor is a depth camera, which can capture the scene depths using ToF of the light. In this chapter, details of digital imaging devices and special sensors for active lighting systems (e.g., thermometers, ToF sensors) are explained.

3.1 Image Sensor

An image sensor (imager) is a sensor which is attached inside a camera to detect the incoming light through a lens and an aperture by converting the light energy into electronic signals. Depending on a physical structure of the photoelectric conversion element, there are two major types (i.e., CCD or CMOS). At the early stages of image sensor development, CCD was the only option because of manufacturing limitations. Although there are many limitations with CCDs, those problems have been mostly mitigated, making the technology a great success with commercial products development. On the other hand, CMOS has not been used for many practical applications since its invention, because primarily CMOS are difficult to produce and they suffer low resolution and high noise. In the late twentieth century, CMOS has emerged in the market with remaining those drawbacks, however, those problems are drastically solved day by day, and finally the image sensor market was dominated by CMOS in 2010, and after as shown in Fig. 3.1.

In the following subsections, basic attributes and features of image sensors are explained. Then, details of CCD and CMOS are provided.

© Springer Nature Switzerland AG 2020
K. Ikeuchi et al., *Active Lighting and Its Application for Computer Vision*,
Advances in Computer Vision and Pattern Recognition,
https://doi.org/10.1007/978-3-030-56577-0_3

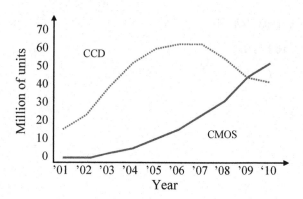

Fig. 3.1 CCD versus CMOS market-share graph for worldwide shipments of image sensors in digital still cameras. Source from IC Insights, WaferNews

3.1.1 Sensor Quality

There are three key factors that define the quality of the image sensors.

Resolution

One of the most important factors of image sensing is resolution. The pixel resolution of the first image sensor was just 100×100 pixels. It, of course, increased rapidly, and now the density is around 2000×2000. Note that simply increasing resolution using the same semiconductor die size will cause image degradation, because it decreases each pixel's size, thus reducing the received energy and leading high noise to keep the same brightness by signal amplification.

Noise

Sensor noise is very important to understand. There are several factors, but three of them are key: photon noise (i.e., photon-shot noise); dark current noise (i.e., dark-shot noise); and read-out noise. Photon noise is inevitable and is caused by fluctuation of the photodetector. A common solution is to increase the exposure time, which takes the average noise value toward zero. However, practically, there are severe restrictions on exposure time. Thus, it remains a challenging task to decrease noise under low-light condition, such as night shot. Dark noise is also a random noise, representing a detected signal in the absence of light. Because the value is independent for each pixel, per pixel calibration is useful for reducing it. However, calibration and real-time process for compensation are quite challenging. Read-out noise results from the bias that occurs when an electron traverses an electronic circuit or is amplified at a transistor. All noises can be reduced by lowering the system temperature.

Dynamic range and sensitivity of photodiode

Among the several elements related to dynamic range and sensitivity, quantum efficiency (QE) is most important, and fill factor is the next. QE is determined by the material quality and the manufacturing process and is not easy to improve. Fill factor is the size of the photodetector for each pixel, and it is certainly possible to improve quality by increasing the die size. This is why the image size continues to grow. To achieve high light efficiency, micro-polygon lenses are usually attached to each pixel to concentrate the light into a valid region of the photodetector. Recently, reverse-side exposure CMOSs, which can avoid the occlusion caused by the photodectector's integrated circuit, has achieved significant progress for dynamic range.

3.1.2 Color Filter

Image sensors output one signal per pixel based on the characteristics of the photoelectric conversion element. These are usually defined by the absorption wavelength/form attributes of the electromagnetic sensor material. Because these attributes do not represent the human visual system, a band-pass filter must be placed in front of the surface to adjust. Moreover, to capture a color image, because the human eye has three different types of photodetector cells, three filters are required, which has different absorption wavelength/form for red, green, and blue.

To capture a full-color image at one time, several solutions have been developed. One puts three different filters at each pixel of the sensor. There are several arrangements of the RGB color filters, but usually, one standard pattern is used. This is the Bayer pattern, as shown in Fig. 3.2. Note that the green color pixel is twice as numerous as the others, because the human eye is more sensitive to the green channel. From a raw-capture image using the Bayer pattern, the original colors are calculated via demosaic algorithms. A drawback of the Bayer pattern is that it captures pseudo-color/shapes when there is a high-frequency texture in the scene, as shown in Fig. 3.2b. To avoid pseudo-colors, a low-pass filter is attached in front of the entire image sensor. Because the filter blurs the image and results in lower resolution output, there is a trade-off between pseudo-colors and blurring with the Bayer pattern. Recently, owing to the achievement of extremely high-resolution image sensors, there is no need for low pass.

Another simple solution captures full-color image using a special prism to optically split colors into three different directions for each image detection. An important advantage of this implementation (i.e., 3CCD/3CMOS) is that every pixel has three colors, rendering the demosaic process unnecessary. Therefore, no low-pass filter is required. However, the high-precision alignment required for the three sensors with pixel level accuracy for the entire image is difficult, especially for high resolution.

3.1.3 Dynamic Range

To record real brightness values for all pixels of a captured image, a wider dynamic range than that of a single image sensor is commonly required. For example, Fig. 3.3, shows an example of a series of real scene captured by the same imager with different exposure time. Among three images of different exposure time, meaningful pixel

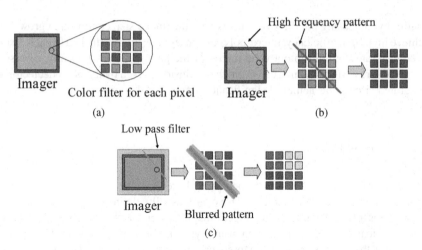

Fig. 3.2 Color filter patterns. **a** Bayer pattern. **b** High-frequency scene is captured, but colors and shapes are not correctly recovered. **c** With low-pass filter, which can recover the pattern

Fig. 3.3 Example of multiple images captured by different exposures. They are integrated to make a single HDRI image

brightness values are only captured in one or two, but are saturated in other images. The relationship between the real and measured pixel values of each image is called response function and the graph is shown in Fig. 3.4. This graph shows a γ curves, which is one of the well-known response curves, and calibration is required to retrieve their γ values. As can be seen, because of the limitations of the sensor's dynamic range, measured pixel values are confined to certain ranges. By fusing several images of different exposure times into one image and adjusting pixel values using a response curve, high-dynamic-range images (HDRI) can be retrieved.

To capture HDRIs, there are several techniques. One typical solution entails capturing multiple images of different exposure times and aperture sizes, fusing them using their respective data. Another solution is putting several filters of different transparencies at neighboring pixels [1].

Fig. 3.4 Response curves of image sensor. It is sometimes intentionally distorted to adjust to a display/projector dynamic range, where response curves are usually nonlinear

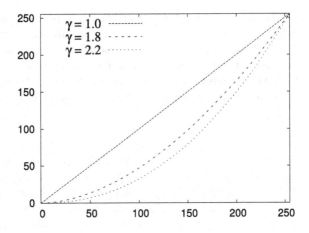

3.1.4   Type of Image Sensor

Based on the attributes and features explained above, details of CCD and CMOS are explained next.

CCD

CCD was first developed by AT&T Bell Labs in 1969, and the first digital camera was developed by Kodak in 1975. Shortly after, Sony, Fuji film and other companies started developing high-resolution CCD digital cameras, which are still sold worldwide. The structure of a CCD is shown in Fig. 3.5. As can be seen, its unique feature is its read-out procedure, in which all electrons in pixels along the same line are sequentially transferred to the circuit gate in a bucket-relay manner. Then, it switches to the next line until all lines are readout. Because all charged electrons are flooded into the attached circuit channels simultaneously, the system is called *global shutter*.

CCD

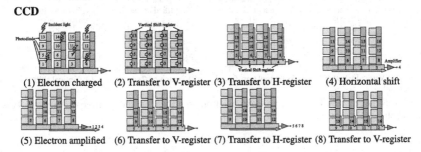

(1) Electron charged   (2) Transfer to V-register   (3) Transfer to H-register   (4) Horizontal shift

(5) Electron amplified   (6) Transfer to V-register   (7) Transfer to H-register   (8) Transfer to V-register

Fig. 3.5 Structure of a CCD sensor and its bucket-relay mechanism. Note that transferring from photodiode to V-register is conducted simultaneously, achieving global shutter on CCD

There are several advantages with CCDs, including high sensitivity, low noise, and global shuttering, which are key reasons for their enduring presence, especially for professional purposes (e.g., broadcasting, industrial inspections, and medical imaging). However, CCD has several severe drawbacks, such as light flooding and high-energy consumption, which are mainly derived from the bucket-relay mechanism, this often makes it more difficult to create high-resolution images than CMOS.

CMOS

CMOS was commercialized several decades after the CCD market. The structure of CMOS is shown in Fig. 3.6. As can be seen, unlike the simpler CCD structure, each CMOS pixel has its own amplifier and other electronic devices attached to it. This causes several problems, such as nonuniformity of pixels, because each pixel has its own parameters for noise, sensitivity, etc. Another problem is the low sensitivity caused by the photodetector being occluded by electronic circuit modules. Such drawbacks have caused practical problems with CMOS application. However, those problems were eventually solved. CMOS increased its pixel resolution drastically because of rapid progress with dynamic random-access memory (DRAM), where the manufacturing process of both silicon semiconductor chips is basically the same. Because of such breakthroughs, the image sensor market has recently been overtaken by CMOS, although CMOS is essentially weak on dynamic ranging because of electronic circuit modules obscuring on the detector. Recently, reverse CMOS has been developed as a solution. Another problem is the rolling shutter effect shown in Fig. 3.7. This is caused by the CMOS read-out being pixel independent and sequentially conducted line-by-line, as shown in Fig. 3.6. Several techniques are being developed to handle this.

CMOS

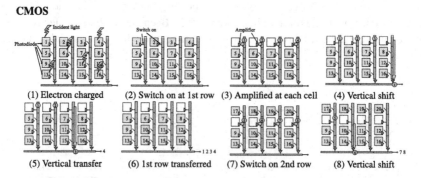

Fig. 3.6 Structure of a CMOS sensor and electron amplification and transferring process. This causes rolling shutter effect

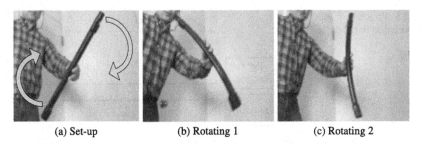

| (a) Set-up | (b) Rotating 1 | (c) Rotating 2 |

Fig. 3.7 Rolling-shutter effect. **a** A stiff and straight bar is rotated and captured by CMOS camera.
b and **c** The bar is apparently curved in captured images during rotation

3.2 Depth Sensor

Because we are living in a 3D space, it is a natural desire for human to capture 3D information on our environment. One simple solution is to apply stereo, which is commonly conducted by most animals including humans using two eyes. Since it is a difficult task to retrieve dense and accurate correspondences between two images, 3D capture with stereo camera is still a challenging task and it has been an important research topic for computer vision. To mitigate the correspondence problem, several types of structured light have been invented and used to encode spatial information into space, which is precisely explained in Sect. 2.5.

Another approach is to directly measure the travel time between a sensor and a target object. For this purpose, sound was first used, followed by electromagnetic wave. Recently light has been used for this purpose, because of its accuracy. Among the light-based sensor systems, ToF is a typical solution.

3.2.1 Principle of Depth Measurement Based on ToF

Because the speed of light is around 300,000 km/s in a vacuum, optical phenomena, such as reflections and refractions, occur seemingly instantly. Therefore, neither the human eye nor a typical camera can capture the propagation of light. Nevertheless, there is a slight difference between the time when light is emitted from a source and when it is reflected by an object's surface to be observed by a sensor. This time difference is the time-of-flight (ToF), and, if it is known, the distance to the object can be estimated.

Let us consider a case in which the light source and sensor are at the same position, as shown in Fig. 3.8. If the time between the light source being turned on is $t = 0$ and that of when the light reaches the sensor after reflection is $t = \Delta t$, the distance d to the object is calculated as

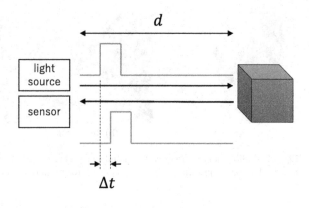

Fig. 3.8 Principle of depth measurement based on ToF

$$d = \frac{\Delta t c}{2}, \qquad (3.1)$$

where c is the speed of light ($c = 2.99792458 \times 10^8 \mathrm{m/s}$).

The depth image can be obtained by calculating the distance for each pixel using an image sensor. Therefore, imaging devices based on ToF are called **ToF cameras** or **depth cameras**.

Compared with various distance estimation methods used in the field of computer vision, there are several advantages. Because system configuration is simple, it is easy to reduce costs and size. No complicated calculations are required, thus distance images at high video rates can be obtained. Because textures are not required, it can be used in various scenes. Additionally, by arranging several light sources around a camera lens, the occlusion problem can be reduced.

3.2.2 Measurement of ToF

Light travels about 30 cm in 1 ns. To estimate the distance with a resolution finer than 30 cm, it is necessary to measure ToF at sub-nanosecond orders. ToF measurement methods are classified into direct and indirect ToF.

Direct ToF

In the direct ToF method, pulse light is emitted from a laser diode, and the time until the reflected light returns is measured directly using a high-speed stopwatch. The Cyrax sensor described in Chap. 11, is categorized as direct ToF. Because the light source emits a very short pulse of light, the quanta of light are extremely small. Therefore, a single-photon avalanche diode (SPAD) is used as the sensor. The SPAD is a type of avalanche photodiode operated in Geiger-mode and used for photon counting.

When a photon enters the SPAD, the electrons are accelerated, and a large avalanche current is generated via avalanche multiplication. Devices, such as time-to-digital converters are required to measure the arrival time of the resulting voltage pulse with a resolution of $10ps$. Additionally, the arrival time of one photon has a large fluctuation. Therefore, the measurement of photons reaching the SPAD is repeated many times to generate a histogram, and the ToF is estimated from the peak position.

SPAD is used as a light detector for light detection and ranging (LiDAR). Scanning LiDAR combines a single SPAD and a scanning mechanism (e.g., polygon mirror). On the other hand, line sensors in which multiple SPADs are arranged 1D and area sensors in which multiple SPADs are arranged 2D have been developed, and are used as flash LiDARs that do not have mechanical parts.

Indirect ToF

In the indirect ToF method, ToF is calculated from the amount of charge (i.e., pixel brightness) measured at the imaging device. Thus, a device having multiple charge storages for each pixel is used instead of a normal imaging device. This is a multi-tap pixel, and the amount of charge is recorded per storage unit per time gate. This device is a photo mixing device (PMD). The Z+F sensor described in Chap. 11 and the Kinect-v2 are categorized as indirect ToF.

The illumination light is temporally modulated, for which there are roughly two types of modulation: short-pulse (SP) and continuous-wave (CW) modulation.

SP modulation illuminates a short pulse of light, and the time delay of the reflected light is calculated from the brightness ratio at each gate. Figure 3.9a shows an example of the simplest 2-tap pixel.

Two gates (i.e., gate-1 and gate-2) are used, and the gate time is T_P. Assuming that the charge amount at each gate is C_1 and C_2, depth d is obtained by

$$d = \frac{cT_P}{2} \frac{C_2}{C_1 + C_2}.$$ (3.2)

Unfortunately, this simple 2-tap pixel configuration is affected by ambient light.

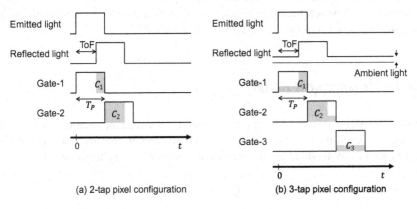

(a) 2-tap pixel configuration   (b) 3-tap pixel configuration

Fig. 3.9 Pixel configurations

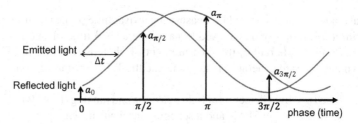

Fig. 3.10 Four-point sampling of the CW modulated light

Figure 3.9b shows an example of a 3-tap pixel configuration. Charge amounts corresponding to gates-1, -2, and -3 are C_1, C_2, and C_3, respectively. Thus, depth d is obtained by

$$d = \frac{cT_P}{2} \frac{C_2 - C_3}{C_1 + C_2 - 2C_3}. \tag{3.3}$$

In this case, the effect of ambient light is corrected by subtracting C_3, which includes only ambient light. In addition to the three gates, a drain is added to discard the charge when the pulsed light does not reach.

CW modulation detects the phase difference of reflected light to the modulated illumination via sinusoidal or rectangular waves. Figure 3.10 shows the principle of the phase detection of modulated light by sampling four points for one period of a sinusoidal wave. Assuming that four sampled signals of every $\pi/2$ are a_0, $a_{\pi/2}$, a_π, and $a_{3\pi/2}$, the optical delay time Δt can be obtained by

$$\Delta t = \frac{1}{2\pi f} \arctan\left(\frac{a_0 - a_\pi}{a_{\pi/2} - a_{3\pi/2}}\right), \tag{3.4}$$

where f is the frequency of the sinusoidal wave. With four-point sampling, the phase can be obtained without being affected by ambient light.

Because the indirect ToF includes calculation processing, it is inferior in accuracy to direct ToF. However, because the pixel configuration is simple, the resolution can be easily increased compared with direct ToF using SPAD.

3.2.3   Measurement Range

With indirect ToF measurement, there is a trade-off between measurement range and accuracy. Thus, if the measurement range is lengthened, the accuracy will decrease. In the case of CW modulation, the maximum distance d_{max} is determined by

$$d_{max} = \frac{c}{2f}. \tag{3.5}$$

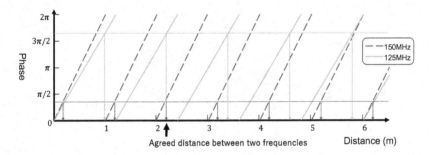

Fig. 3.11 Phase unwrapping

Because the phase is defined from 0 to 2π, when it exceeds 2π, it returns to 0 again. This problem is called phase wrapping. For example, when the frequency is 150 MHz, d_{max} becomes 1 m. Thus, distances of 1 m or longer cannot be measured. Ambiguity occurs in 1-m increments. In other words, when the distance measurement result is 0.2 m, it is not known if the actual distance is 0.2, 1.2 m, or 2.2 m.

The phase-wrapping problem can be easily resolved by combining multiple frequencies. Figure 3.11 shows an example when 150 MHz and 125 MHz are combined. They produce wrapping at 1.0 and 1.2 m, respectively. Because the wrapping periods are different, the distances that match at both frequencies is correct. In this example, wrapping occurs every 6.0 m, but the measurement range is expanded while maintaining accuracy.

3.2.4 Multi-path Problem

Until now, we have assumed only a simple optical path wherein the light emitted from a light source is reflected on the object's surface and observed by a sensor. In reality, there are more complicated paths, as shown in Fig. 3.12. If light from multiple optical paths is observed together, the distance cannot be measured correctly. This problem is called the multi-path problem.

Typical causes of multi-path problems are interreflection and scattering. Interreflection is a phenomenon in which light reflected from one surface illuminates another. The brightness value becomes significant when other objects are nearby. Therefore, concave shapes cannot be measured well. However, scattering is caused by repeated collisions with fine particles. This phenomenon is divided into volume and subsurface scattering. Because volume scattering occurs in fog, object distances are often incorrectly measured closer than reality. Because subsurface scattering occurs on translucent plastics, the estimated distances are farther than reality.

In the case of SP modulation, the reflected light of the rectangular pulse from a light source is mixed and greatly distorted. Therefore, the distance calculations of Eqs. (3.2) and (3.3) fail. However, with CW modulation, if the illumination is

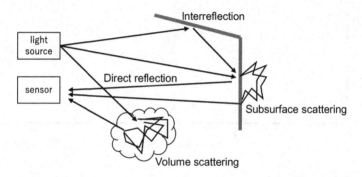

Fig. 3.12 Multi-path problem

modulated using a sinusoidal wave, it will remain a sinusoidal wave, even if the reflected light is mixed. However, the phase does not correspond to distance. Because the distance measurement by ToF is performed pixel-wise, a global optical path must be considered to solve the multi-path problem. In the case of CW modulation, it is known that, by increasing the frequency, reflected multipath light cancels each other and the error of the distance measurement becomes smaller.

3.3   Other Sensors

3.3.1   Hyperspectral Camera

This section introduces cameras that can obtain the multispectral images of a scene. Multispectral refers to wideband 4–10-channel images/cameras, and hyperspectral refers to narrowband 20–1,000-channel images/cameras with 1–20-nm full width at half-maximum spectral sensitivity with symmetrical shapes arranged at equal intervals. Most people do not care about the different attributes associated with these two terms, and this book does not strictly distinguish their meanings. Both terms are commonly used for four or more channels used to distinguish common color images, because those images/cameras are three-channel (RGB). However, *multispectral* is sometimes used for two or more channels to distinguish them from single channels when needed.

The six-band multispectral camera developed by Tominaga et al. [2] comprises a single RGB camera and two filters (Fig. 3.13). They calculated a six-band multi-spectral image using an image with one filter and an image with another filter set in front of the camera. These filters were multispectral filters and were designed so that six-channel images could be obtained when combined with an RGB camera.

A beam-splitting multispectral camera is illustrated in Fig. 3.14. The light path can be split using a combination of prisms if they are adequately configured to operate

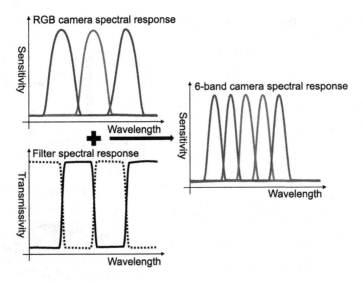

Fig. 3.13 Six-channel multispectral camera

Fig. 3.14 Beam-splitting multispectral camera

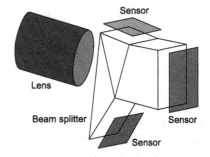

as a beam splitter. The FD-1665 (FluxData, Inc., NY USA) is the beam-splitting multispectral camera that allows custom-made configurations. It has a three-way beam splitter and three imaging sensors. Combinations of monochrome/color sensors and color filters result in multispectral camera with three-to-nine bands.

VariSpec filter (PerkinElmer, Inc., MA USA) is a filter which passes the light of intended wavelength. VariSpec filter uses liquid crystal which is electronically controllable. A series of optical elements are bounded together in this filter. The filter works like a narrowband band-pass filter with intended wavelength. Multispectral image can be obtained if we set this filter in front of a monochrome camera. Since the light incoming to the camera is weak due to the narrow wavelength, noise-suppressed camera such as cooled CCD camera is required.

Dispersing elements, such as prisms or diffraction gratings, are used to split the light into discrete wavelengths (Fig. 3.15). Many commercial multispectral cameras use such elements (e.g., NH series (Eba Japan Co., Ltd., Japan)). If a 1D slit of light passes through the dispersing element, the transmitted light spreads in 2D. To capture

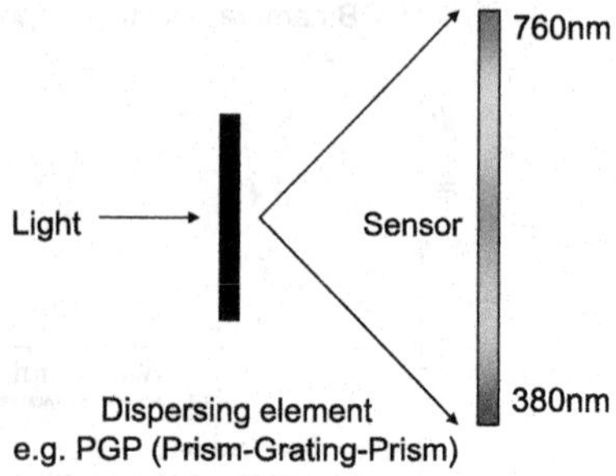

Fig. 3.15 Push-broom-type
multispectral camera

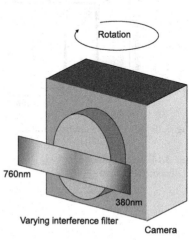

Fig. 3.16 Panorama
multispectral camera

the multispectral image of a 2D scene, these cameras usually scans the observing area in one direction. The scanned line is then dispersed into the 2D image. Thus, if we scan the scene in a direction vertical to the dispersion direction, we obtain a 2D multispectral image of the scene.

Multispectral images captured by the system developed by Schechner and Nayar [3] are panoramic (Fig. 3.16). This capturing system uses a varying interference filter, which passes light of a specific wavelength. However, the wavelength that is transmitted depends on the spatial area of the filter. The wavelength that passes through this filter gradually differs from one discrete wavelength to another from one end of the filter to the other. This filter is set in front of the camera, and the camera is rotated horizontally in small steps. The captured image sequence is integrated, resulting in a panoramic multispectral image.

The multispectral device developed by the Interuniversity Microelectronics Centre, Belgium, is a 16- or 25-band multispectral camera (Fig. 3.17). The image sensor is arranged by sets of 4×4 or 5×5 sensors with different spectral sensitivities. One block has 16 or 25 sensor chips and such blocks are arranged over the entire surface to form an imaging sensor. Demosaicking the multiband sensor using different wavelength results in a multispectral image having 16 or 25 channels.

Fig. 3.17 Mosaic-type
multispectral camera

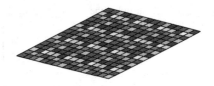

Applications of multispectral analysis has widely spread. Some of photometric stereo methods use multispectral light. Material identification, for example, can be performed via multispectral analysis (Sect. 8.1). One example of novel multispectral light applications is shown in Sect. 9.1.7.

3.3.2 Polarization Camera

The simplest polarization camera is one with the linear polarizer mounted in front of it. A linear polarizer is often expressed as a circle filled with straight lines (Fig. 3.18a). The orientation of the line expresses the orientation of the light oscillation to be transmitted through the polarizer. Note that the orientation of the real structure is orthogonal for both dichroic linear polarizers and wire-grid linear polarizers (Fig. 3.18b). The iodine molecule of a dichroic linear polarizer and the metal molecules of a wire-grid linear polarizer are aligned orthogonally to the oscillating orientation of the transmitted light.

When measuring the polarization state of light using a camera with a linear polarizer, the camera gamma should be 1. Please be careful not to create a gap between the linear polarizer and lens. When rotating the filter around the optical axis of the camera direction, the brightness of the image will change. Three or more images having different rotating angles of the linear polarizer will enable the calculation of the parameter of linear polarization. If we rotate the linear polarizer, the brightness changes sinusoidally as follows (Fig. 3.19), because the period of linear polarizer is 180°.

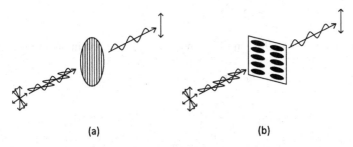

(a)   (b)

Fig. 3.18 Linear polarizer. **a** Illustration. **b** Dichroic structure

Fig. 3.19 Sinusoid of the brightness while rotating the linear polarizer

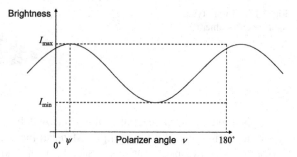

$$I = \frac{I_{\max} + I_{\min}}{2} + \frac{I_{\max} - I_{\min}}{2} \cos(2\nu - 2\psi) . \tag{3.6}$$

Here, $I_{\max}$ and $I_{\min}$ are the maximum and the minimum values of the sinusoid, respectively. The rotating angle of the linear polarizer is denoted as ν. The angle where $I_{\max}$ is observed is called the phase angle, ψ. Some papers use an angle wherein $I_{\min}$ is observed as a phase angle ψ. Equation (3.6) is also represented as follows:

$$I = a \sin \beta + b \cos \beta + c , \tag{3.7}$$

where

$$a \sin \beta + b \cos \beta = \sqrt{a^2 + b^2} \sin(\beta + \alpha) ,$$

$$\sin \alpha = \frac{b}{\sqrt{a^2 + b^2}} ,$$

$$\cos \alpha = \frac{a}{\sqrt{a^2 + b^2}} ,$$

$$c = \frac{I_{\max} + I_{\min}}{2} ,$$

$$\sqrt{a^2 + b^2} = \frac{I_{\max} - I_{\min}}{2} ,$$

$$\beta = 2\nu ,$$

$$\alpha = \pi/2 - 2\psi .$$

If we substitute $0°$, $45°$, $90°$, and $135°$ into Eq. (3.6), we obtain the following:

$$I_0 = \frac{I_{\max} + I_{\min}}{2} + \frac{I_{\max} - I_{\min}}{2} \cos(2\psi) , \tag{3.8}$$

$$I_{45} = \frac{I_{\max} + I_{\min}}{2} + \frac{I_{\max} - I_{\min}}{2} \sin(2\psi) , \tag{3.9}$$

$$I_{90} = \frac{I_{\max} + I_{\min}}{2} - \frac{I_{\max} - I_{\min}}{2} \cos(2\psi) , \tag{3.10}$$

$$I_{135} = \frac{I_{max} + I_{min}}{2} - \frac{I_{max} - I_{min}}{2} \sin(2\psi). \tag{3.11}$$

The degree of polarization, ρ, for linear polarization is defined as follows:

$$\rho = \frac{I_{max} - I_{min}}{I_{max} + I_{min}}. \tag{3.12}$$

If we capture four images with $0°$, $45°$, $90°$, and $135°$ angles of linear polarization, the degree of polarization and the phase angle can be calculated as follows:

$$\rho = \frac{\sqrt{(I_0 - I_{90})^2 + (I_{45} - I_{135})^2}}{(I_0 + I_{45} + I_{90} + I_{135})/2}, \tag{3.13}$$

$$\psi = \frac{1}{2}\text{atan2}(I_{45} - I_{135}, I_0 - I_{90}). \tag{3.14}$$

Here, atan2 represents the arctangent function.

$$\text{atan2}(y, x) = \tan^{-1}\frac{y}{x}. \tag{3.15}$$

If we capture three images with $0°$, $45°$, and $90°$ angles of linear polarization, the degree of polarization and the phase angle can be calculated as follows:

$$\rho = \frac{\sqrt{(I_0 - I_{90})^2 + (2I_{45} - I_0 - I_{90})^2}}{I_0 + I_{90}}, \tag{3.16}$$

$$\psi = \frac{1}{2}\text{atan2}(2I_{45} - I_0 - I_{90}, I_0 - I_{90}). \tag{3.17}$$

If we capture three or more images with $\nu_1, \nu_2, ..., \nu_N$ angles of linear polarization, the polarization parameters, a, b, and c, shown in Eq. (3.7), can be calculated from the pixel brightness values, $I_1, I_2, ..., I_N$.

$$\begin{pmatrix} I_1 \\ I_2 \\ \vdots \\ I_N \end{pmatrix} = \begin{pmatrix} \sin\beta_1 & \cos\beta_1 & 1 \\ \sin\beta_2 & \cos\beta_2 & 1 \\ \vdots & \vdots & \vdots \\ \sin\beta_N & \cos\beta_N & 1 \end{pmatrix} \begin{pmatrix} a \\ b \\ c \end{pmatrix}. \tag{3.18}$$

The closed form solution can be obtained from the known angles of linear polarizers $\nu_1, \nu_2, ..., \nu_N$, where $\beta_1 = 2\nu_1$, $\beta_2 = 2\nu_2$, ..., $\beta_N = 2\nu_N$ hold.

Schechner [4] estimated both polarization parameters and angles of linear polarization under the condition that the angle of the linear polarizer in Eq. (3.18), was unknown. His method used multiple pixels having different polarization parameters (a, b, and c in Eq. (3.18)). He proved that, if there were two pixels having different polarization parameters, all parameters could be obtained from the captured images

Fig. 3.20 Twisted nematic liquid crystal

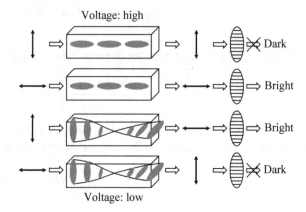

of five or more different angles of linear polarization. Additionally, he proved that if there were three or more pixels having different polarization parameters, all parameters could be obtained from four or more angles of linear polarization.

Wang et al. [5] used an LCD (liquid crystal display) in order to calculate the angles of linear polarization. LCD provides perfectly polarized light, and can show a checkerboard pattern. The method was robust, because the geometrical calibration could be performed using the checkerboard pattern, and it could use multiple data at multiple viewpoints. They not only estimated the angles of linear polarization, they also estimated the camera response function.

Wolff et al. [6], Fujikake et al. [7], Harnett and Craighead [8] developed a system to obtain the polarization image by using a liquid crystal, which can twist the angle of polarization, and it is electronically controllable (Fig. 3.20). The light first passes through the liquid crystal. Then, it passes through a fixed linear polarizer, and, finally, it reaches the camera. From the relationship between the brightness and the twisting parameter of liquid crystal, we can estimate the angle and the degree of polarization.

Lanthanum-modified lead zirconate titanate (PLZT) is a transparent ferroelectric ceramic. PLZT is a birefringence medium, and it acts like a retarder (waveplate). The optical properties of PLZT depend on the electric field. By changing the amount of field, PLZT will act like a half-wave plate, a quarter-wave plate, etc. Polarization cameras that use liquid crystal only measure the first three components of the Stokes vector. A PLZT polarization camera can measure all four components [9].

Photonic Lattice, Inc., Japan, developed a polarization camera whose CCD had four kinds of linear polarizers with its size 5 μm in front of it (Fig. 3.21). Therefore, this polarization camera could measure the polarization state of a scene in real time. This kind of polarization camera uses assorted pixels (Fig. 3.21) and is now available from various manufacturers. The Phoenix polarization camera (LUCID Vision Labs, Inc., Canada) uses a SONY Polarsens CMOS imaging sensor [10], which detects four types of polarization states of light at 0°, 45°, 90°, and 135° angles of linear polarization.

Fig. 3.21 Assorted pixels
for a polarization camera

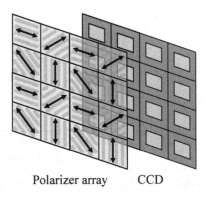

Polarizer array   CCD

FIR cameras and FIR linear polarizers are necessary to analyze the polarization
states of FIR light. FIR should be observed at 4- or 10-μm wavelengths, because light
at other wavelengths is absorbed by the atmosphere. These limitations are referred
to as the *atmospheric window*. Specac Limited, UK, and Edmund Optics, NJ USA,
provide FIR linear polarizers.

3.3.3 Thermography/Thermal Imaging Camera

Infrared is not visible to human eye, however, important for active lighting systems
because of its special features, e.g., they do not cause interference with human visual
systems, and they penetrate fog and haze very well. There are five sub-regions of
infrared as shown in Table 3.1. Because the shorter wavelengths of NIR slightly
overlap visible light, Si-based sensors can detect the corresponding wavelengths by
simply changing the RGB filter to the appropriate wavelength without changing the
image sensor itself. Because Si chip cannot detect wavelength more than 1000 nm,
another sensor material or a totally different mechanism is required for measurement.

Table 3.1 Infrared sub-division by wavelength

Name	Wavelength	Frequency (THz)
Near-infrared (NIR)	780–1400 nm	214–380
Short-wavelength infrared (SWIR)	1.4–3 μm	100–214
Mid-wavelength infrared (MWIR)	3–8 μm	37–100
Long-wavelength infrared (LWIR)	8–15 μm	20–37
Far-infrared (FIR)	15–1000 μm	0.3–20

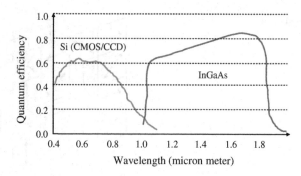

Fig. 3.22 Sensitivity feature of Si and InGaAs

For SWIR, because InGaAs has a sensitive bandwidth range of 900–1700 nm, as shown in Fig. 3.22, InGaAs-based sensor has been developed and commercial products are available. However, because InGaAs fabrication has not yet been streamlined, it is difficult to achieve the high resolution of Si chips. Furthermore, the cost of InGaAs chips is still very high, because they are made of rare metals. Nonetheless, recent development of InGaAs chip is intensively conducted, because SWIR has high transparency through fog and haze and has drawn a great deal of attention from the industry for its autonomous driving applications. Notably, long-range LiDAR for outdoor environments is strongly demanded for safe driving.

For longer wavelengths, bolometers are used for wave measurement. This device measures temperature changes by precisely measuring the resistance change of electricity, as shown in Fig. 3.23. Light is absorbed by the blackened light-receiving surface, and its temperature rise is measured using a resistance thermometer. Thus, there is no wavelength selectivity in theory. A material having a large temperature coefficient of resistance is selected. For metal, a platinum strip having a thickness of ∼0.1 μm (blackened surface) or a deposited film of nickel, bismuth, antimony (Sb), etc., (thickness 0.05–0.1 μm) is used. Because FIR conveys heat energy efficiently farther away, it is possible to measure the wave by selecting an appropriate bandpath filter for the corresponding wavelength. To measure a high-resolution 2D area

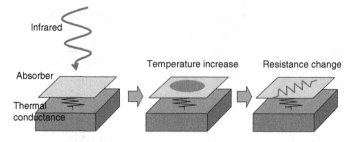

Fig. 3.23 Bolometer mechanism

for FIR, a micro-bolometer has been developed to be arranged in a smaller space, achieving greater than 640×480 commercial resolution.

3.3.4 Interferometry

If electromagnetic waves are split in two by a prism or slits, they traverse different paths prior to being merged again after interfering with each other. Using sensors to measure this kind of interference is called *interferometry*. Figure 3.24 shows a standard configuration of Michelson interferometry [11]. As shown, if the path lengths of two split waves are the same when merged, the phases are also the same, and the brightness increases with interference. However, if the two phases are opposite, the brightness becomes darker. Because such brightness difference comes via different path lengths, by observing their variations, it is possible to detect the difference of lengths between the two paths, providing clues to object depth with extremely high accuracy, such as the wavelength of light (500–700 nm for visible light).

Among the wide range of electromagnetic waves, laser is ideal for measuring the depth of an object with high accuracy, because it produces coherent light and the efficiency of interference is high: optimal for optical interferometry. To capture the 2D region, scanning a scene by a line laser using Galvano or MEMS mirrors has been developed. With such a configuration, stripe patterns are typically observed, as shown in Fig. 3.25, based on the length differences at each point. By using laser-beam expanders, it is possible to observe 2D areas of an object's surface at once. Because the wavelength of a common laser is between 500–800 nm, we gain the ability to detect differences of two depths with orders of accuracy.

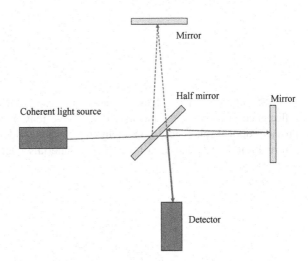

Fig. 3.24 Configuration of interferometry

Fig. 3.25 Optical
interferometry

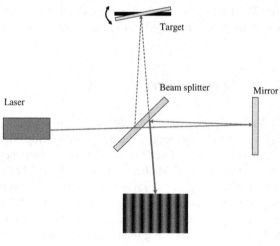

Fig. 3.26 Optical
Coherence Tomography

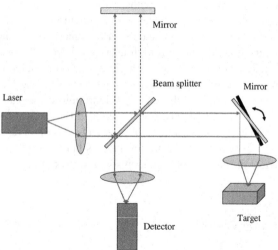

By using a condensing lens, as shown in Fig. 3.26, it is possible to observe the
reflectance of a specific depth by changing the lens' focal length. Such a system
provides optical coherence tomography (OCT) and is widely used to measure volume
information near the surface of translucent materials, such as human pupils.

3.3.5 BRDF Measurement Sensor

As explained in Sect. 1.3, the surface reflectance of an opaque object's surface can be defined using four parameters, including view direction (θ_v, ϕ_v) and light direction (θ_l, ϕ_l), as shown in Fig. 1.8. Thus, the relationship between light's input and output at a specific point on an object's surface can be described using a 4D BRDF function. Although BRDF maps to 4D space, and because the surface reflectance of a real object is usually smooth in terms of viewpoint and light-source positional changes, the function is normally expressed using simple models having few parameters. The Lambert model, a typical reflectance model, is represented using one parameter (i.e., albedo). A change of intensity is then explained using the cosine function. Specularity is another type of reflectance and several efficient models have been proposed, all applying a small number of parameters with simple equations [12]. However, considering the existence of the huge number of real objects in the world, there are many whose BRDF cannot be represented with a simple model or a small number of parameters. Thus, capturing the actual reflectance value of a full 4D BRDF remains an important and challenging task. To efficiently capture BRDF, many systems have been proposed and developed. In the early stages of BRDF measurement, the typical system consisted of a camera, a light source, and one or two robotic arms, which precisely controlled the camera and/or light source positions [13]. Its drawback was long scanning times, because the 4D measurement space is extremely large, and sampling a single value in a single capture is inefficient. Therefore, a subset of 4D BRDF (e.g., 2D or 3D) is normally captured. To efficiently decrease dimensions, the symmetric nature of BRDF is utilized. For example, if an angle between the surface normal and the half-vector remains unchanged, the light intensity is invariant to the view direction.

Fig. 3.27 Photometric sampler developed by Nayar et al. [14]

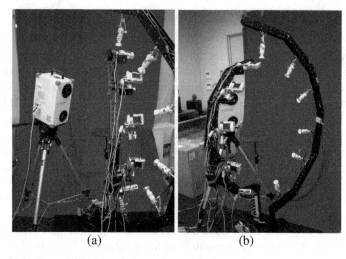

Fig. 3.28 BTF image capturing platform [15]. **a** Equipped with range sensor. **b** Concentric arc system

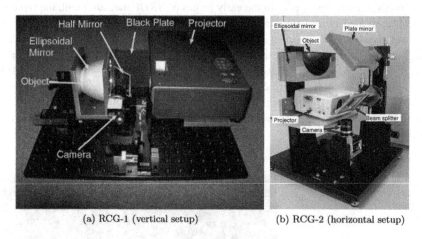

(a) RCG-1 (vertical setup)   (b) RCG-2 (horizontal setup)

Fig. 3.29 Ellipsoidal mirror-based sampling machine developed by Mukaigawa et al. [16]

In terms of a real system being used to capture BRDF, Nayar et al.proposed a photometric sampler that could capture 2D BRDF from a single viewpoint with a discrete 2D light position distributed across a hemisphere [14], as shown in Fig. 3.27. Furukawa et al.proposed a full bi-directional texture function (BTF)-capturing device that used multiple cameras and multiple light sources to achieve real-time photo-realistic rendering [15] as shown in Fig. 3.28. The system required calibration of all the cameras and light positions, however, capturing efficiency and precision was extremely high. Mukaigawa et al.proposed a system that used an ellipsoidal mirror to capture 2D BRDF at a single shot as shown in Fig. 3.29. They also used a video projector, which illuminated only a single dot in a single capture and changed its pixel position to cover the entire screen to capture 3D BRDF without any mechanical devices [16]. Ben–Ezra et al.proposed a system that measured the BRDF of a full hemisphere by using an LED as both a light emitter and a detector [17]. In their method, because no camera device was required, the BRDFs of a full hemisphere or sphere could be acquired without occlusions.

References

1. Nayar SK, Mitsunaga T (2000) High dynamic range imaging: Spatially varying pixel exposures. In: Proceedings of the IEEE conference on computer vision and pattern recognition (CVPR), vol 1. IEEE, pp 472–479
2. Fukuda T, Kimachi A, Tominaga S (2008) A high-resolution imaging system for omnidirectional illuminant estimation. J Imaging Sci Technol 52(4):40907-1
3. Schechner YY, Nayar SK (2002) Generalized mosaicing: wide field of view multispectral imaging. IEEE Trans Pattern Anal Mach Intell (PAMI) 24(10):1334–1348
4. Schechner YY (2015) Self-calibrating imaging polarimetry. In: 2015 IEEE international conference on computational photography (ICCP). IEEE, pp 1–10
5. Wang Z, Zheng Y, Chuang YY (2019) Polarimetric camera calibration using an LCD monitor. In: Proceedings of the IEEE conference on computer vision and pattern recognition (CVPR), pp 3743–3752
6. Wolff LB, Mancini TA, Pouliquen P, Andreou AG (1997) Liquid crystal polarization camera. IEEE Trans Robot Autom 13(2):195–203
7. Fujikake H, Takizawa K, Aida T, Kikuchi H, Fujii T, Kawakita M (1998) Electrically-controllable liquid crystal polarizing filter for eliminating reflected light. Opt Rev 5(2):93–98
8. Harnett CK, Craighead HG (2002) Liquid-crystal micropolarizer array for polarization-difference imaging. Appl Opt 41(7):1291–1296
9. Miyazaki D, Takashima N, Yoshida A, Harashima E, Ikeuchi K (2005) Polarization-based shape estimation of transparent objects by using raytracing and plzt camera. In: Polarization science and remote sensing II, vol 5888. International society for optics and photonics, p 588801
10. Yamazaki T, Maruyama Y, Uesaka Y, Nakamura M, Matoba Y, Terada T, Komori K, Ohba Y, Arakawa S, Hirasawa Y et al (2016) Four-directional pixel-wise polarization cmos image sensor using air-gap wire grid on 2.5-μm back-illuminated pixels. In: 2016 IEEE international electron devices meeting (IEDM). IEEE, pp 8-7
11. Michelson AA, Morley EW (1887) On the relative motion of the earth and of the luminiferous ether. In: Sidereal messenger, vol 6, pp 306–310
12. Phong BT (1975) Illumination for computer generated pictures. Commun ACM 18(6):311–317
13. Stanford Graphics Lab (2002) Stanford spherical gantry. https://graphics.stanford.edu/projects/gantry/
14. Nayar SK, Sanderson AC, Weiss LE, Simon DA (1990) Specular surface inspection using structured highlight and gaussian images. IEEE Trans Robot Autom 6(2):208–218
15. Furukawa R, Kawasaki H, Ikeuchi K, Sakauchi M (2002) Appearance based object modeling using texture database: acquisition compression and rendering. In: Rendering techniques, pp 257–266
16. Mukaigawa Y, Sumino K, Yagi Y (2009) Rapid brdf measurement using an ellipsoidal mirror and a projector. IPSJ Trans Comput Vis Appl 1:21–32. https://doi.org/10.2197/ipsjtcva.1.21
17. Ben-Ezra M, Wang J, Wilburn B, Li X, Ma L (2008) An led-only BRDF measurement device. In: Proceedings of the IEEE conference on computer vision and pattern recognition (CVPR). IEEE, pp 1–8

Chapter 4
Light Source

To build an active-lighting system, selection and configuration of light sources are crucial. In this chapter, we categorize existing light sources based on several criteria and explain the principles and mechanisms of light emission as well as photometric and geometric features of several typical light sources.

As explained in Chap. 1, visible light comprises electromagnetic waves with wavelengths between 380 and 760 nm. Light sources can be characterized by wavelength, brightness, polarization, etc. Basically, light sources can be divided into two categories: natural and artificial. Natural light sources include the sun, flames, lightning, bioluminescence, and more. Artificial lights include light bulbs, fluorescent lamps, pattern projectors, LEDs, lasers, and more. Although natural light sources are important for computer vision and are sometimes used as active light, they are not common for active-lighting techniques, because natural light sources cannot be precisely controlled. Thus, we only consider artificial light sources in this chapter.

We categorize artificial light sources into several types based on criteria such as (1) wavelength, (2) emission principle, and (3) light source model and distribution, as shown in Fig. 4.1.

4.1 Light Emission Principle and Mechanism

There is a long history of the controversy about whether light is a wave or a particle. Of course, we now know that light possesses both properties until measured, and does, in fact, comprise electromagnetic waves as described in Chap. 1. The creation of light has long been a challenge for humans. For more than 10,000 years until about 150 years ago, bonfires and candle flames were the key methods of creating artificial light. Things changed rapidly with the invention of the light bulb. Furthermore, LED is a more recent type and has drastically overtaken artificial light-source

© Springer Nature Switzerland AG 2020
K. Ikeuchi et al., *Active Lighting and Its Application for Computer Vision*,
Advances in Computer Vision and Pattern Recognition,
https://doi.org/10.1007/978-3-030-56577-0_4

Principle	Distribution Wave length		Point				Line	Area
			Point (0D)	Line (1D)	Area (2D) High freq. (Pattern)	Area (2D) Low freq. /Directional		
Black body	B r o a d	visible		Boundary of cast shadow	Video projector	Sun/ Lamps (Incandescent, Halogen)		Sky/ Indirect lighting
		NIR				Sun		Sky
		FIR				Sun/ Heat lamp		Sky
Fluorescence	P e a k s	UV					Black light	
		visible					Fluorescent lamp/ Neon tube	
Band-gap		visible	Point laser	Line laser	LED projector/ Laser projector/ VCSEL			
		NIR	Lidar light source		TOF light source/ Laser pattern projector			
		FIR	Quantum cascaded laser					

Fig. 4.1 Light sources classified by several categories

production because of its energy efficiency, compactness, long life, wavelength variation, etc. In the following subsections, the principles for light emission mechanisms are explained.

4.1.1 Blackbody Radiation

If an object is heated, its temperature rises. Temperature refers to the energy level of an atom or a molecule. Higher temperature creates stronger atomic vibration. When the velocity of atoms and molecules decreases, energy is lost typically in the form of light quanta (photon). When the temperature reaches an equilibrium condition where the object eventually returns to its original temperature by losing its excess energy via thermal conduction, radiation, light emission, etc., it is called blackbody radiation. This type of radiation only depends on an object's temperature. Blackbody radiation energies are defined by the following equation:

$$E = nh\nu (n = 0, 1, 2, ...),\tag{4.1}$$

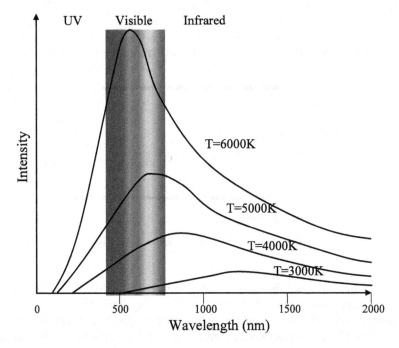

Fig. 4.2 Wavelength of blackbody radiation

where ν is oscillation frequency, and $h = 6.626 \times 10^{-34}[J \cdot s]$ is Planck's constant. Wavelength and energy examples at several temperatures are shown in Fig. 4.2. Via the quantized photon release when objects lose energy, lower temperature objects emit reddish colors. This color phase extends to blue and violet when the temperature rises. A typical blackbody radiator is the sun. Its temperature ranges around 5,777 K, and its peak spectrum is 460 nm, which is same as green. This has caused all earthly life forms to evolve with green color sensitivity.

4.1.2 Fluorescence and Phosphorescence

Fluorescence was first observed by Sir John Frederick (UK) at about ∼450 nm of blue light in 1845. Fluorescence is the property of receiving light and re-emitting it at longer wavelengths, as shown in Fig. 4.3. Quinine, for example, is excited by blue light to produce yellow fluorescence. Using this as a filter for wine, we can block blue scattered light and observe only its yellow fluorescence.

It is known that when a fluorescent component is added to an object's color, it is very clear and bright. This is because receiving light energy is excited and re-emitted, and the process is very short, generally in the order of nanoseconds. Also, when the received light energy is re-emitted, emitted wavelength is longer than that

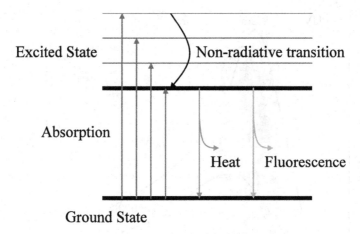

Fig. 4.3 Principle of fluorescence based on energy absorption and emission

of received light, because the energy of long wavelength is lower than that of short wavelength, and energy cannot be increased without any other extra inputs. In detail, the absorption of light excites electrons in the molecule, and excess energy is emitted as longer wavelength photons when electrons from the first electron excitation surface return to the ground state through various processes. Furthermore, light emission from electrons transitioning to an electronically excited triplet and returning to their ground state is called *phosphorescence*.

4.1.3 LED

LED is a semiconductor and, unlike conventional incandescent and fluorescent lamps, they produce light using a mechanism in which electrical energy changes directly into light in a semiconductor crystal as shown in Fig. 4.4. This phenomenon was first discovered in 1907. However, it took a long time before LEDs were invented by Nick Holonyak in 1962. At the time of their invention, LEDs were only red. A yellow-green LED was later invented in 1972 by George Crawford. In the early 1990s, Shuji Nakamura, Isamu Akasaki, and Hiroshi Amano invented a blue LED semiconductor using Ga nitride, winning 2014's Nobel prize. In recent years, improvements have been remarkable; and white LEDs with the same luminous efficiency as fluorescent lamps now exist.

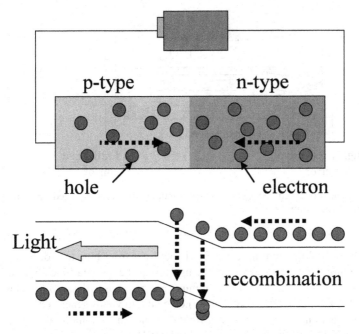

Fig. 4.4 Principle of LED. Light emission occurs when protons and electrons are merged

4.1.4 Laser

Laser is also a semiconductor to produce light by directly converting electric energy into light. The most important feature of a laser is its high beam coherence, because it can extract light with precisely the same properties (e.g., wavelength and phase). Such laser oscillation is achieved by amplifying the light generated by injecting current that reciprocates between two mirrors as shown in Fig. 4.5. In simple terms, a laser diode is an LED that emits light by amplifying light with a reflector.

4.1.5 Bioluminescence

Bioluminescence is a phenomenon in which living organisms generate and emit light. This occurs as a result of a chemical reaction that converts chemical energy into light energy. Bioluminescence basically emits light when luciferin is oxidized and the excited oxide (oxyluciferin) becomes a fluorescent light emitter when it returns to its ground state. Luciferin varies depending on the luminescent organism, and eight types have been identified thus far, including the well-known D-Luciferin (i.e., firefly).

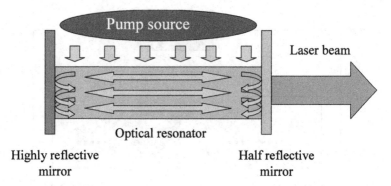

Fig. 4.5 Light is reflected between two mirrors until the light power exceeds the threshold and emits through the mirror

The feature object luminescence is visible light from blue to red. For example, marine luminescent creatures hide themselves in the sea via blue light, and land fireflies use yellow-green light to communicate between males and females. Additionally, the light generation efficiency (i.e., quantum yield) generated from a single chemical reaction is high. This *cold light* does not involve thermal radiation. The quantum yield of firefly emission is said to be 0.41.

4.2   Geometric Model and Distribution Characteristics

Generally, most light sources are represented using a combination of **light-source models** and **light distribution characteristics**. For example, a halogen lamp is usually approximated as a point light source. However, light distribution is not uniform, and, even the same point source is observed from the same distance; the light is strongest in the direction directly in front of the lamp. The spotlight is another example of extreme light distribution in one direction (i.e., directional light).

In terms of a **light-source model**, the most basic one is a point light source. In this model, light is emitted from a point into a 3D hemisphere. The brightness decays in inverse proportion to the square of the distance from the light source. It is notable that many light sources, such as (halogen) light bulbs, and candlelight, are categorized as point light sources. Moreover, if a light distribution is given as a 2D pattern, it is also possible to explain video projectors as a point light source. Another common light source model is a linear one in which point light sources are lined up. A fluorescent bulb is a typical example of a linear light source. Another important light source is an area light source in which point light sources are arranged in 2D over an object surface. This model is particularly useful in the research field of computer graphics to represent global light illumination for photo-realistic rendering.

Light distribution characteristics are important, because many light sources do not emit light uniformly; the light brightness differs depending on the direction of the

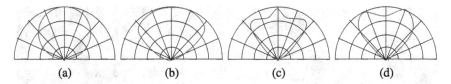

Fig. 4.6 Example of nonuniform light distribution for a point light source. **a** Circle-shaped. **b** Petal-shaped. **c** Bell-shaped. **d** Heart-shaped

light source. Usually, lamps for industrial or professional purposes have specification sheets, as shown in Fig. 4.6. There are typical shapes for these distributions, including bell-shaped, double-ring-shaped, or heart-shaped.

4.2.1 Light Source Geometric Model

Point Light Source

Incandescent lamps and candles are typical examples of a point light source. Halogen light is generated from a type of incandescent bulb and is used when a long lifetime and a high luminance light source are required. Because of such attributes, a halogen lamp has commonly been used as a light source for LCD projectors. However, when compared with ideal point light sources, (1) the light distribution characteristic of incandescent bulbs is uneven, and (2) the size of the light source cannot be ignored.

Active-lighting systems often use a light source in which a special light distribution is artificially provided as a projection pattern. Let us consider giving a special light distribution to a point light source. In this case, the light distribution is given by a parameter of the directional vector from the source using a spherical function. Thus, an ideal laser beam can be thought of as a point light source with a light distribution characteristic defined on a unit sphere such that the function value corresponding to one point (e.g., $(x, y, z) = (0, 0, 1)$) is 1, and for the other direction is 0. An ideal line laser (laser sheet in 3D space) is another example that provides light distribution characteristics such that the function value corresponding to a plane (e.g., $x = 0$) passing through the origin and the intersection of unit spheres is 1. Video projectors can also be defined by light distribution characteristics in a 2D space on a unit sphere of a point light source.

Linear Light Source

Fluorescent light a typical linear light source, because the shape of the bulb is usually straight and linear. There is also a circle-shaped fluorescent bulb, which is also categorized as a linear light source. Another example is a neon tube, as shown in Fig. 4.7. Although neon tubes are linear light sources, they are usually curved and bent to represent letters or figures.

Fig. 4.7 Neon tube as an example of a linear light source. Image by Shutterbug75 from Pixabay

Because linear light sources can be considered infinitely connected point light sources, they create soft shadows because the distance between occluders and shadow-casting objects is far apart.

Area Light Source

If point or linear light sources are covered by a diffuser, the entire body of the diffuser object glows. These are *area light sources*. Another example is intentional indirect lighting generated by illuminating ceilings or walls with point or linear light sources where the light reflects back into the living space. The sky is an area light source, wherein the atmosphere is a huge diffuser to the sun.

Because area light sources illuminate scenes from various directions, it is collectively called *global illumination* and makes soft shadows behind occluders. Shadow boundaries are usually softer than those of linear light sources.

4.2.2 Light Distribution Characteristics

Light distribution can be divided into two categories, high and low frequency. High-frequency distribution includes patterns of dots, lines, curves, and their combinations. The distribution is frequently used for structured light techniques.

On the other hand, low-frequency distributions are also important for active-lighting systems. For example, a uniform distribution to a single direction is used for photometric stereo techniques. Because the actual light distribution of a point light source is not uniform (e.g., bell- or heart-shaped), those distributions must be calibrated in advance and compensated. Additionally, some specific low-frequency patterns (e.g., sinusoidal) are widely used for structured light in conjunction with phase analyses.

Although there are several geometric light source models, as explained in the previous section, point light sources are usually adopted for 2D pattern projection because of their simple optical structures, easy pattern designs, and proliferation.

0D (point) Projection

Point projection is a light source having a light distribution characteristic of only one point on the hemisphere, where one area is lit and the other area is not. Laser pointers are well-known examples. A point laser takes advantage of this property to generate light from a narrow range of light sources that can be focused to approximately one point and can be regarded substantially as rays. However, the opening gate from which light is emitted is not necessarily a point, and in many cases, it has a certain size even if it is extremely narrow.

1D (line) Pattern Projection

The line pattern is a 1D line, and is realized by causing a point pattern, such as a laser pointer, to be spread in a planar manner using a cylindrical lens or the like. When such a light plane is reflected on the surface of an object, it is observed as a line or a curve, which can be considered a collection of illuminated points on the plane.

Because line lasers extend along a plane in 3D space, they are often modeled as planes in many applications, such as light sectioning. However, because the actual light source is small, areas that cannot be seen by the light source exist via self-occlusion and are not illuminated.

2D (area) Pattern Projection

A point laser produces a 0D (point) pattern, and a line laser produces a 1D (line) pattern. However, video projectors and some laser projectors can project a 2D pattern from one point into space. In the sense of projecting a pattern from one point, point and line lasers are the same. However, by projecting a 2D pattern, the surface of a 3D object can be scanned by single projection, whereas multiple projections are required for point or linear projections. For example, when using a point laser, it is necessary to scan 2 DoF in the x and y directions to obtain a 2D depth image. When using a line laser, a 1-DoF scan is required to a different direction from that of the line of the line laser (typically vertically).

Because the information obtained by the camera is 2D, a light source projecting a 2D pattern can obtain a large amount of information from just a single capture. By devising 2D patterns, it can be applied to various applications, and it has a high affinity with 3D shape measurements. Therefore, it is often called a projector–camera system (e.g., Procams). A particularly practical system using a projector–camera system obtains depth data by applying triangulation between a projector and a camera. If the 2D pattern of the projector and the corresponding points of the 2D image captured by the camera can be acquired, the wide-range shape on the image screen can be measured at once.

As explained, with depth measurement, the acquisition of corresponding points is critical. Similarly, with projection mapping and other techniques, the correspondences between the pixels of the projector and the points on the 3D scene is important. This can be modeled as an inverse camera, and, in fact, the same projection model as a camera is usually used for projector–camera configurations. Details are explained in Sect. 2.5.

2D (area) Directional (parallel) Projection

A uniform pattern can be considered a typical example of a low-frequency pattern. This is because, when observing light from a point light source from a very distant place, all the lights have approximately the same direction. In such limited situations, all light rays have the same direction vector. A light source that emits such a light is called a directional light. A typical example of a directional light is sunlight. A parallel beam projector in which point light sources are formed by topocentric optics is also a directional light.

4.3   Light Sources

Details of general light sources commonly used for active-lighting systems are explained.

4.3.1   Incandescent Lamp

Incandescent light bulbs have been used for more than a century. In the case of a transparent bulb, the actual light source is part of a filament. If the filament is small enough for the working space, this can be considered a point light source. Additionally, filament lamps are often regarded as point light sources with uniform light distribution characteristics, because their brightness variation in several directions is relatively small. Bamboo was once used as the filament, but tungsten is used now.

4.3.2   Halogen Lamp

A halogen lamp containing a halogen element is used when a long lifetime and a high luminance value or brightness are required. It can produce a sufficiently even light distribution with enough brightness via liquid crystals. Thus, it is frequently used as a light source for LCD projectors. For the same reason, it is also used for digital micro-mirror devices (DMD) and other video projectors. In many real packages, the light distribution is often biased in one direction by a reflector or the like to increase light efficiency. Therefore, uniform light distribution characteristics are not rigorously correct for actual systems.

4.3.3 Fluorescent Lamp

As described with the principle of fluorescence, a fluorescent lamp is one that irradiates a fluorescent substance with an ultraviolet light generated by an electric discharge within the lamp, extracting visible light as a re-emission. Ultraviolet rays are generated when electrons emitted from an anode to a cathode through a vacuum chamber contact mercury gas inside a fluorescent lamp. The mechanism of the fluorescent lamp is that the ultraviolet light reacts with the fluorescent substance to produce visible light.

To emit electrons into the lamp, *emitters* are attached to both ends of the fluorescent tube. Additionally, an inert gas is enclosed in the fluorescent lamp with mercury gas, and the pressure is reduced to 2–4 hPa. The inner surface of the tube is coated with a fluorescent substance that emits light in response to ultraviolet light. For commercial products, there are several white-color lights based upon blackbody principles, such as bulb color, warm white color, and daylight color lights. These are used as attributes for home and commercial use, depending on the type of materials. For example, for the light of a family in a living room, bulb color is suitable. For a study room, daylight color is suitable. The features of the fluorescent lamps are as follows:

1. Owing to long life and high lamp efficiency, they are economically efficient.
2. Construction of large light-emitting areas achieving smaller shadow regions is possible.
3. High frequency types have been developed and used achieving quick response and high efficiency.

There are three types of lighting methods for fluorescent lamps: glow starter, rapid start, and inverter (for high-frequency lighting).

1. Glow-starter type
   This method uses a glow starter (i.e., lighting tube). Because the circuit is simple and is generally cheap, it is widely used. A disadvantage is that it takes time from switch-on to lighting and induces flickering.
2. Rapid-start type
   This type is designed to light up instantly. However, the ballast is large and heavy. This type was mainstream until it was overtaken by energy-saving inverter types.
3. Inverter type
   The ballast of the inverter type comprises electronic circuits. It is more efficient, lighter, and can be turned on instantly. It operates at a high frequency. Thus, there

is no flicker, and the brightness per watt is high. Because a high frequency is generated after returning the alternating current to direct current, it can be used, regardless of the power-supply AC-frequency value.

4.3.4 Black Light (Ultraviolet Light Source)

Black light is a type of fluorescent lamp used for appraisal and special lighting effects. This lamp cuts visible light to a ~352 nm wavelength, which is in the near-ultraviolet region, providing a strong fluorescent reaction. Special filters that absorb visible light and transmit ultraviolet light are coated.

The main purpose of the light is for appraisal and appreciation of documents and ores, and detection of stains and dirt on clothes. It is also used as special lighting effects for stage and signboards. Although these lights can be attached to ordinary fluorescent lighting equipment, most reflectors on those chassis are coated with white melamine, which barely reflects light of a wavelength <400 nm. Therefore, it is effective to use an aluminum (Al) reflector. Usually, black light wavelengths do not include a bactericidal effect. However, it can adversely affect eyes and skin if exposed for a long time. Therefore, it is advisable to wear glasses and gloves for protection.

4.3.5 LED

The structure of an LED is a combination of a semiconductor P-type electrode and an N-type electrode. When electricity flows, electrons flow from the N-type to the P-type, and light is emitted when two streams collide. LED emissions include white, red, blue, and green light, and the principle is that the color changes depending on the compound used in the LED chip. Compounds used in these semiconductors include Ga, nitrogen (N), In, Al, phosphorus (P), germanium (Ge), Si, and arsenic (As). Depending on the compound, different wavelengths of light are emitted, changing the color of the LED. There are two ways to create white light: mixing two or more complementary colors or mixing the three RGB primary colors. Many colors can be expressed through these combinations, and brighter illuminations have increased. Recently, an LED that emits near-ultraviolet light (wavelength 380–420 nm) has been developed. Using this as an excitation light source, a white LED that can emit the entire visible light range, such as a fluorescent lamp, has been created.

Compared to fluorescent and incandescent lamps, LED bulbs have the following advantages, and thus, most light sources are rapidly replaced by LEDs.

1. Low power consumption
   Similarly bright fluorescent lamps typically consume 7 W and incandescent bulbs typically consume 40 W, whereas LED bulbs consume only 5 W or less.

2. Long lifetime

Bulb-type fluorescent lamps last up to 10,000 h, whereas incandescent bulbs last for ~2000 h. LEDs last for approximately 40,000 h, four times longer than fluorescents.

3. Running cost

Fluorescent lamps and LEDs are more cost-effective than incandescent bulbs, but initial costs are higher. If one incandescent bulb's price is $1, a fluorescent is $5, and an LED is $50. However, considering the power consumption efficiency and long life, the LED investment is superior.

4. Brightness

Fluorescent and incandescent lamps are often displayed in terms of power consumption W (watts) as a guide, whereas LED bulbs are displayed in lm (lumen), which indicates brightness. Incandescent light bulbs and fluorescent lights spread light in all directions. However LEDs have directional distributions, which are suitable for lighting efficiency and designing lighting environments.

5. Size

Because the LED has a sink for heat dissipation, the metal parts other than the light emitting part are larger. This is because the LED working components are sensitive to heat. Thus, the size of the bulb is often the same or larger for LEDs.

6. Light color

Unlike incandescent or fluorescent lamps, LED bulbs allow users to choose their favorite light colors according to the scene by combining several spectral LEDs into a single bulb.

4.3.6 Laser

A laser light source creates coherent light by utilizing properties, such as resonant light emission using a semiconductor, an excited gas, and similar ones. Laser light is significantly polarized around a certain wavelength in its frequency distribution and maintains strong energy over a long distance. Based on these features, laser is used as a light sources for various active-lighting systems, such as LiDAR with frequency modulation, video projectors with micro-electromechanical systems (MEMS), ToF with a beam expander, speckle techniques, diffractive optical elements (DOE), and many more.

4.3.7 Vertical Cavity Surface-Emitting Laser (VCSEL)

Semiconductor lasers generally resonate light in a direction parallel to the substrate surface, emitting light in that direction. VCSEL uses a high-reflection distributed Bragg reflector (DBR), providing a stacked structure of semiconductors or dielectrics as a reflecting mirror to resonate light in a direction perpendicular to the substrate surface and to emit the light in a direction perpendicular to the surface.

Owing to its unique structure, beam quality and temperature characteristics are reasonable, and it can be driven by low power consumption and high-speed modulation is possible. Because of such advantages, VCSEL is frequently used as the light source for ToF applications.

4.3.8 Video Projector

The typical video projector uses liquid crystal. This allows for real-time changes in projection pattern and has often been used for shape measurement. A commercially available projector has a large aperture compared to a camera. Thus, it also has a shallow depth of field near its focus depth. Additionally, the light of the pattern is attenuated by the inverse square of the distance.

Another representative example of a video projector is a digital light-processing (DLP) projector. This type controls the color with a color wheel while controlling the direction and activation (on-off) of each pixel with a MEMS element, as shown in Fig. 4.8. For this reason, if it is observed in a short period of time, only a semi-random 2D pattern distribution is seen. Such properties can be used for ultra-high-speed measurement [1].

There are also devices that realize a 2D expression by scanning a laser with a DMD or galvanic mirror.

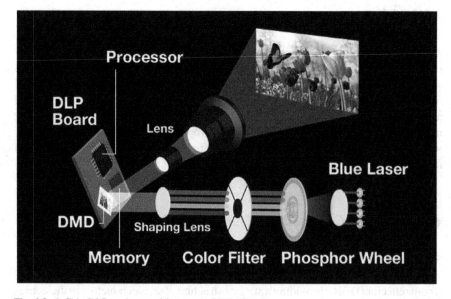

Fig. 4.8 1-Chip DLP projector of Panasonic PT-RZ575

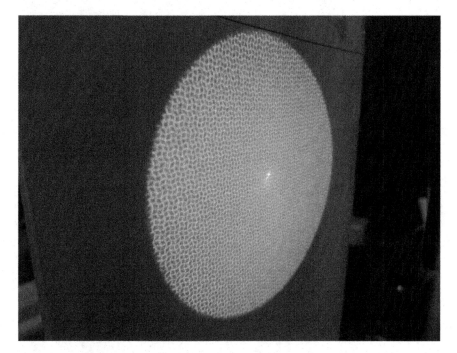

Fig. 4.9 DOE pattern projection using green laser as a light source

4.3.9 DOE Laser Projector

The DOE element splits a point laser using diffraction. The light after branching also has the property of laser light, so the pattern does not spread. Attenuation is small and it also has the ability of phase interference. Figure 4.9 shows an example of a DOE pattern. It should be noted that there is an inevitable single strong dot called zeroth-order light at the center of the pattern. Because of high-energy efficiency, DOE is frequently used for projector–camera-based structured light systems [2, 3].

References

1. Narasimhan SG, Koppal SJ, Yamazaki S (2008) Temporal dithering of illumination for fast active vision. In: Proceedings of the European conference on computer vision (ECCV). Springer, pp 830–844
2. Microsoft (2010) Xbox 360 Kinect. http://www.xbox.com/en-US/kinect
3. Furukawa R, Sanomura Y, Tanaka S, Yoshida S, Sagawa R, Visentini-Scarzanella M, Kawasaki H (2016) 3d endoscope system using doe projector. In: 2016 38th annual international conference of the IEEE engineering in medicine and biology society (EMBC). IEEE, pp 2091–2094

Part II
Algorithm

Chapter 5
Photometric Stereo

Shading on a surface provides strong cues for us to perceive the 3D shape of an object. The study of computationally inferring a 3D shape from shading cues originated with *shape from shading* [1, 2], which aims to estimate a 3D shape from a single image observed under a light source. The problem has attracted many researchers' attentions since then because of its mathematically rich problem structure and important practical applications. Later, it was shown that, instead of a single image, the use of multiple images observed under different lighting conditions could alleviate the difficulty of the problem of 3D shape estimation from shading cues. This multiple-image approach is known as *photometric stereo*. This chapter describes the problem of photometric stereo and its solution methods.

5.1 Photometric Stereo Problem

Photometric stereo is a class of methods for determining the shape of a static scene from a set of observations recorded from a fixed viewpoint but under varying lighting conditions. A shape estimate is obtained in the form of a *surface normal*, which corresponds to the surface gradient expressing the surface orientation. More formally, surface normal is represented as a three-dimensional unit vector $\mathbf{n} \in \mathcal{S}^2 (\subset \mathbb{R}^3)$,[1] which relates to the continuous surface as a height map $z(x, y) : \mathbb{R}^2 \to \mathbb{R}$ viewed from a camera as

$$\mathbf{n} = \frac{\left[-\frac{\partial z}{\partial x}, -\frac{\partial z}{\partial y}, 1 \right]^\top}{\left\| \left[-\frac{\partial z}{\partial x}, -\frac{\partial z}{\partial y}, 1 \right] \right\|_2} \tag{5.1}$$

---

[1] $\mathcal{S}^2 = \{ \mathbf{v} \in \mathbb{R}^3 : \|\mathbf{v}\|_2 = 1 \}$.

© Springer Nature Switzerland AG 2020
K. Ikeuchi et al., *Active Lighting and Its Application for Computer Vision*,
Advances in Computer Vision and Pattern Recognition,
https://doi.org/10.1007/978-3-030-56577-0_5

in xyz-world coordinates. We begin with describing the image formation model, i.e., how a pixel brightness measurement is produced via the interaction of the light, surface normal, and surface reflectance. To make the explanation simple, we assume a grayscale input in this section.

A basic photometric stereo method is built upon the *Lambertian* image formation model , written as

$$m \propto \rho E \mathbf{l}^\top \mathbf{n}, \tag{5.2}$$

where $m \in \mathbb{R}$, $m \geq 0$ is a measured brightness at a scene point corresponding to the pixel of interest, $\mathbf{l} \in \mathcal{S}^2$ is a light direction vector, $\rho \in \mathbb{R}$, $\rho \geq 0$ is the Lambertian diffuse albedo, and $E \in \mathbb{R}$, $E > 0$ is the brightness of the light source. Since only from the measurement m it is impossible to separate the magnitudes of albedo ρ and light brightness E, and the global scaling of the radiance to measured brightness is unknown, with a slight abuse of notation, we denote the Lambertian image formation model hereafter as

$$m = \rho \mathbf{l}^\top \mathbf{n}. \tag{5.3}$$

In this manner, diffuse albedo ρ includes the global scaling and the light brightness E. Besides, we use $\tilde{\mathbf{n}} = \rho \mathbf{n} \in \mathbb{R}^3$ to represent a surface normal $\mathbf{n}$ scaled by diffuse albedo ρ. With this notation, Eq. (5.3) can be simplified to

$$m = \mathbf{l}^\top \tilde{\mathbf{n}}. \tag{5.4}$$

In what follows, we assume that a scene is illuminated by a directional light unless it is stated otherwise. We also assume an orthographic projection model of a camera, whose viewing direction is oriented along $[0, 0, 1]$ pointing toward the origin, and that image coordinates are aligned with the world's x-y coordinates.

Let us extend the image formation model of Eq. (5.4) to multiple scene points. With p scene points (corresponding to p pixels) observed under a single light direction $\mathbf{l}$, we have

$$\begin{bmatrix} m_1 \ m_2 \ \dots, m_p \end{bmatrix} = \mathbf{l}^\top \begin{bmatrix} \tilde{\mathbf{n}}_1 \ \tilde{\mathbf{n}}_2, \dots, \tilde{\mathbf{n}}_p \end{bmatrix}. \tag{5.5}$$

Given f images taken under varying light directions $\{\mathbf{l}^1, \dots, \mathbf{l}^f\}$, it can be written in a matrix form as

$$\underbrace{\begin{bmatrix} m_1^1 \ m_2^1 \ \dots, m_p^1 \\ m_1^2 \ m_2^2 \ \dots, m_p^2 \\ \vdots \ \vdots \ \ddots \ \vdots \\ m_1^f \ m_2^f \ \dots, m_p^f \end{bmatrix}}_{\mathbf{M} \in \mathbb{R}^{f \times p}} = \underbrace{\begin{bmatrix} \mathbf{l}^{1\top} \\ \mathbf{l}^{2\top} \\ \vdots \\ \mathbf{l}^{f\top} \end{bmatrix}}_{\mathbf{L} \in \mathbb{R}^{f \times 3}} \underbrace{\begin{bmatrix} \tilde{\mathbf{n}}_1 \ \tilde{\mathbf{n}}_2, \dots, \tilde{\mathbf{n}}_p \end{bmatrix}}_{\mathbf{N} \in \mathbb{R}^{3 \times p}}, \tag{5.6}$$

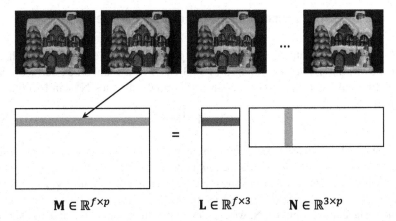

Fig. 5.1 Matrix form of the Lambertian image formation model. An observed image is vectorized to form a row vector in the measurement matrix $\mathbf{M}$. The light matrix $\mathbf{L}$ contains light vectors (row vectors) corresponding to the observed images. The surface normal matrix $\mathbf{N}$ has an albedo-scaled surface normal vector (column vector) at each pixel location

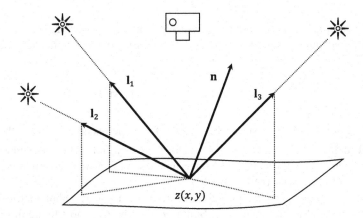

Fig. 5.2 Photometric stereo determines the surface normal $\mathbf{n}$ at each scene point (x, y, z) on a surface $z(x, y)$ from a set of measurements observed under different lighting conditions $\{\mathbf{l}_1, \mathbf{l}_2, \ldots\}$

resulting in a linear system of equations (see Fig. 5.1) in which the matrix of surface normal $\mathbf{N}$ is unknown:

$$\mathbf{M} = \mathbf{LN}. \tag{5.7}$$

The goal of photometric stereo is to determine the surface normal $\mathbf{N}$ from the measured pixel brightnesses $\mathbf{M}$ obtained under different lighting conditions $\mathbf{L}$ as depicted in Fig. 5.2. In a *calibrated* photometric stereo setting, the light matrix $\mathbf{L}$ is given, while *uncalibrated* photometric stereo that treats the light matrix $\mathbf{L}$ is also unknown. In what follows, we mainly discuss the calibrated setting, where the light matrix $\mathbf{L}$ is known unless it is stated otherwise.

5.2 Lambertian Least Squares Photometric Stereo

In 1980, Woodham introduced the original photometric stereo method [3]. It has
been shown that given three images ($f = 3$ case) of a Lambertian surface taken
under different light directions, the surface normal estimates $\mathbf{N}^*$ can be obtained
from Eq. (5.7) by

$$\mathbf{N}^* = \mathbf{L}^{-1}\mathbf{M} \tag{5.8}$$

if the inverse $\mathbf{L}^{-1}$ exists. In case the light matrix $\mathbf{L}$ does not have its inverse (rank
deficient case), the solution cannot be uniquely determined. Such a degenerate case
occurs when the light vectors $\mathbf{l}^1$, $\mathbf{l}^2$, and $\mathbf{l}^3$ are coplanar or even colinear, corresponding
to rank $(\mathbf{L}) = 2$ and rank $(\mathbf{L}) = 1$ cases, respectively. Once we obtain the solution
for the surface normal matrix $\mathbf{N}$, we can derive the diffuse albedo ρ and unit surface
normal vector $\mathbf{n}$ for each pixel by

$$\rho = \|\tilde{\mathbf{n}}\|_2, \quad \mathbf{n} = \frac{\tilde{\mathbf{n}}}{\rho} \tag{5.9}$$

from the definition that the surface normal vector $\mathbf{n}$ is a unit vector, i.e., $\|\mathbf{n}\|_2 = 1$.

When given more than three images ($f > 3$ case), we may obtain a least squares
approximate solution of surface normal $\mathbf{N}^*$ as

$$\mathbf{N}^* = \left(\mathbf{L}^\top \mathbf{L}\right)^{-1} \mathbf{L}^\top \mathbf{M} = \mathbf{L}^\dagger \mathbf{M} \tag{5.10}$$

if the inverse $\left(\mathbf{L}^\top \mathbf{L}\right)^{-1}$ exists. In Eq. (5.10), $\mathbf{L}^\dagger$ is the pseudoinverse of the light
matrix $\mathbf{L}$ defined as $\mathbf{L}^\dagger = \left(\mathbf{L}^\top \mathbf{L}\right)^{-1} \mathbf{L}^\top$. The necessary and sufficient condition for
the inverse to exist is the same as the three-image setting, namely, the light matrix $\mathbf{L}$
needs to be non-singular (rank $(\mathbf{L}) = 3$).

Figure 5.3 shows an example of Lambertian least squares photometric stereo
applied to a set of input images of a Bunny scene. The rightmost figure in Fig. 5.3
shows the estimated surface normal map. This visualization of a surface normal is
commonly used in the study of photometric stereo, in which the surface normal's
xyz directions are encoded in RGB color channels.

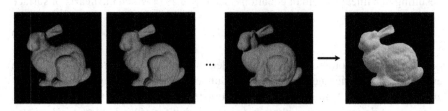

Fig. 5.3 Lambertian least squares photometric stereo. The rightmost figure represents the estimated
surface normal given the input images

5.3 Robust Photometric Stereo

The least squares Lambertian photometric stereo introduced in Sect. 5.2 works well for the Lambertian observations. However, if the reflectances deviate from the Lambertian assumption, they introduce errors in surface normal estimates $\mathbf{N}^*$. The examples of the *non-Lambertian* observations are illustrated in Fig. 5.4.

One of the major sources of deviations from the Lambertian assumption is due to attached shadows. In the image formation model of Eq. (5.3), it is assumed that a scene point is always illuminated by light sources. However, if the surface normal $\mathbf{n}$ does not face toward the light source direction $\mathbf{l}$, i.e., $\mathbf{l}^{\top}\mathbf{n} \leq 0$, the measured pixel brightness m becomes zero. This effect is called attached shadow. By accounting for the attached shadows, Eq. (5.3) is rewritten as

$$m = \rho \max \left(\mathbf{l}^{\top}\mathbf{n}, 0 \right), \tag{5.11}$$

where max operator chooses the maximum value of the arguments. Another frequently observed non-Lambertian observations are due to specular reflections, which are mirror-like reflections. In contrast to diffuse reflection, in which incoming light is scattered and bounced off of the surface in a wide range of directions, incident rays are reflected around the same angle to the surface normal but toward the opposing side of it. Other major sources of errors are due to cast shadows and interreflections.

The Lambertian least squares photometric stereo is fairly simple. However, in reality, Lambertian surfaces rarely (or maybe even never) exist due to the illumination effects stated above, and the Lambertian model can only be used as an approximate

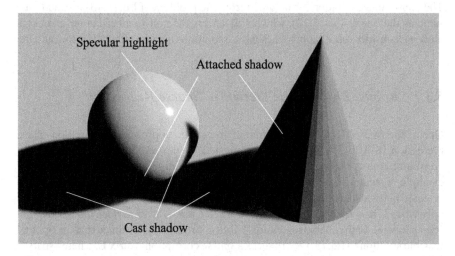

Fig. 5.4 In real-world situations, there are non-Lambertian reflectances, such as specular reflections, cast and attached shadows. Such observations affect the Lambertian least squares photometric stereo method due to the deviation from the Lambertian assumption

reflectance model. In fact, when there are non-Lambertian observations, the accuracy of surface normal estimates $\mathbf{N}$ is affected. These non-Lambertian observations can be treated as *outliers* that do not fit the Lambertian model. We can therefore consider an extended Lambertian model that accounts for such outliers:

$$\mathbf{M} = \mathbf{LN} + \mathbf{E}, \tag{5.12}$$

where $\mathbf{E} \in \mathbb{R}^{f \times p}$ is an additive unknown corruption matrix that contains anything that deviates from the Lambertian model, applied to an otherwise ideal Lambertian observation. There have been a number of *robust* estimation methods to this problem via, e.g., deterministic inlier selection [4, 5], RANSAC [6], and median filtering [7], to name a few. These individual methods have different assumptions over $\mathbf{E}$, but essentially they can be represented by the model shown in Eq. (5.12) where $\mathbf{N}$ and $\mathbf{E}$ are treated as unknowns.

With the explicit model with corruptions $\mathbf{E}$ in Eq. (5.12), the goal becomes to recover the surface normal $\mathbf{N}$ as a part of the Lambertian component $\mathbf{LN}$ in the presence of unknown non-Lambertian corruptions $\mathbf{E}$. However, this is an under-constrained problem since the number of unknowns in $\mathbf{N}$ and $\mathbf{E}$ exceeds the number of linear equations f in Eq. (5.12).

One of the simplest ways to disambiguate the infinitely many solutions to Eq. (5.12) is to rely on a least squares penalty applied to corruption $\mathbf{E}$. It can be written in an energy minimization manner as

$$\mathbf{N}^*, \mathbf{E}^* = \underset{\mathbf{N}, \mathbf{E}}{\operatorname{argmin}} \|\mathbf{E}\|_F^2 \quad \text{s.t.} \quad \mathbf{E} = \mathbf{M} - \mathbf{LN}. \tag{5.13}$$

The approximate solution $\mathbf{N}^*$ obtained from the problem of Eq. (5.13) is exactly the same as the one in Eq. (5.10), which still suffers the non-Lambertian observations. Now we consider robust approaches whose solutions are less affected by the outliers.

5.3.1 Robust Photometric Stereo via Sparse Regression

One of the most effective robust photometric stereo approaches is based on sparse regression [8–10]. The approach is motivated by the reasonable observation that non-Lambertian observations emerge primarily in limited areas of each image. For example, specular highlights only surround the spot where the surface normal is oriented halfway between lighting and viewing directions, while shadows are created only when $\mathbf{l}^\top \mathbf{n} \leq 0$ (attached shadow) or when a non-convex surface blocks the light (cast shadow). In the sparse regression-based approach, it is assumed that the optimal feasible solution to Eq. (5.12) produces a *sparse* corruption matrix $\mathbf{E}$, where "sparse" indicates that most of the matrix elements are zero while there are only a small number of nonzero elements.

With an assumption of sparse corruptions $\mathbf{E}$, we could turn the problem into a sparse regression problem formulated as

$$\mathbf{N}^*, \mathbf{E}^* = \underset{\mathbf{N}, \mathbf{E}}{\operatorname{argmin}} \|\mathbf{E}\|_0 \quad \text{s.t.} \quad \mathbf{E} = \mathbf{M} - \mathbf{LN}. \tag{5.14}$$

The only difference from Eq. (5.13) is that the matrix norm in the objective function is turned into the ℓ_0 norm instead of the Frobenius norm. The ℓ_0 norm is the number of nonzero entries in the matrix; thus, its minimization encourages the sparsity of the corruption matrix $\mathbf{E}$.

Let us now look at a pixel-wise expression of the problem for the purpose of simpler explanation. In the pixel-wise expression, Eq. (5.14) becomes

$$\tilde{\mathbf{n}}^*, \mathbf{e}^* = \underset{\tilde{\mathbf{n}}, \mathbf{e}}{\operatorname{argmin}} \|\mathbf{e}\|_0 \quad \text{s.t.} \quad \mathbf{e} = \mathbf{m} - \mathbf{L}\tilde{\mathbf{n}} \tag{5.15}$$

with $\mathbf{m} = \begin{bmatrix} m^1 & m^2 & \dots, & m^f \end{bmatrix}^\top \in \mathbb{R}^f$ being a measurement vector observed under f lightings and a column error vector $\mathbf{e} \in \mathbb{R}^f$. As in the matrix case, the ℓ_0 norm of a vector corresponds to the number of nonzero elements, i.e., $\|\mathbf{v}\|_0 = \operatorname{nnz}(\mathbf{v})$; thus, the objective function in Eq. (5.15) seeks the solution that makes the error vector $\mathbf{e}$ sparse. This ℓ_0 formulation is preferable over the ℓ_2 norm based one because ℓ_0 norm is unaffected by the magnitude of outliers. However, unfortunately, minimization of an ℓ_0 norm term is difficult because it is highly non-convex and discontinuous.

One viable solution technique is to relax the ℓ_0 norm term $\|\mathbf{e}\|_0$ by a convex ℓ_1 norm term $\|\mathbf{e}\|_1$ as commonly done in the machine learning and statistics literature. With this relaxation, the problem of Eq. (5.15) is approximated to its convex surrogate as

$$\tilde{\mathbf{n}}^*, \mathbf{e}^* = \underset{\tilde{\mathbf{n}}, \mathbf{e}}{\operatorname{argmin}} \|\mathbf{e}\|_1 \quad \text{s.t.} \quad \mathbf{e} = \mathbf{m} - \mathbf{L}\tilde{\mathbf{n}}. \tag{5.16}$$

The ℓ_1 norm approximate solution yields a sparse error vector $\mathbf{e}$, effectively capturing the outliers. As a result, the surface normal estimate $\tilde{\mathbf{n}}^*$ becomes less affected by outliers resulting in a greater accuracy. The use of ℓ_1 norm approximate solution method was first introduced by Wu et al. [8] and later adopted by [9, 10].

In Eq. (5.16), it is assumed that the constraint $\mathbf{e} = \mathbf{m} - \mathbf{L}\tilde{\mathbf{n}}$ exactly holds. If the assumption is untrue due to some factors, such as imperfect light calibration, we can relax the hard constraint by an additional model mismatch penalty as

$$\tilde{\mathbf{n}}^*, \mathbf{e}^* = \underset{\tilde{\mathbf{n}}, \mathbf{e}}{\operatorname{argmin}} \left\{ \|\mathbf{e}\|_1 + \lambda \|\mathbf{m} - \mathbf{L}\tilde{\mathbf{n}} - \mathbf{e}\|_2^2 \right\}, \tag{5.17}$$

where $\lambda \in \mathbb{R}, \lambda \geq 0$ is a nonnegative trade-off hyperparameter balancing data fit with sparsity. For example, in the limit as $\lambda \to \infty$ makes the problem identical to Eq. (5.16).

There are a number of solution techniques for constrained and unconstrained ℓ_1 norm minimization problems, corresponding to Eqs. (5.16) and (5.17), respectively. Here, we briefly describe a linear programming (LP)-based method for the constrained problem Eq. (5.16) and an iteratively reweighted least squares (IRLS) method for the unconstrained problem Eq. (5.17).

Constrained ℓ_1 Norm Minimization Eq. (5.16) via Linear Programming (LP)

It is understood that the constrained ℓ_1 norm minimization problem can be cast to LP [11, Chap. 6]. By stacking unknowns $\tilde{\mathbf{n}}$ and $\mathbf{e}$, Eq. (5.16) can be rewritten as

$$\tilde{\mathbf{n}}^*, \mathbf{e}^* = \underset{\tilde{\mathbf{n}}, \mathbf{e}}{\operatorname{argmin}} \left\| \begin{bmatrix} \mathbf{0} & \mathbf{I} \end{bmatrix} \begin{bmatrix} \tilde{\mathbf{n}} \\ \mathbf{e} \end{bmatrix} \right\|_1 \quad \text{s.t.} \quad \begin{bmatrix} \mathbf{L} & \mathbf{I} \end{bmatrix} \begin{bmatrix} \tilde{\mathbf{n}} \\ \mathbf{e} \end{bmatrix} = \mathbf{m}, \qquad (5.18)$$

where $\mathbf{0}$ is a zero matrix whose elements are all zeros, and $\mathbf{I}$ is an identity matrix, with appropriate sizes. Let us denote

$$\mathbf{x} = \begin{bmatrix} \tilde{\mathbf{n}} \\ \mathbf{e} \end{bmatrix} \in \mathbb{R}^{(3+f)}, \quad \mathbf{C} = \begin{bmatrix} \mathbf{0} & \mathbf{I} \end{bmatrix} \in \mathbb{R}^{f \times (3+f)}, \quad \text{and} \quad \mathbf{B} = \begin{bmatrix} \mathbf{L} & \mathbf{I} \end{bmatrix} \in \mathbb{R}^{f \times (3+f)}.$$

Then, Eq. (5.18) can be written with the new notations above as

$$\mathbf{x}^* = \underset{\mathbf{x}}{\operatorname{argmin}} \| \mathbf{C} \mathbf{x} \|_1 \quad \text{s.t.} \quad \mathbf{B} \mathbf{x} = \mathbf{m}.$$

Let us introduce a new vector variable $\mathbf{t} \in \mathbb{R}^f$ whose ith element t_i is the absolute value of the ith element of $\mathbf{C}\mathbf{x}$. Then the problem becomes

$$\mathbf{x}^*, \mathbf{t}^* = \underset{\mathbf{x}, \mathbf{t}}{\operatorname{argmin}} \mathbf{1}^\top \mathbf{t} \quad \text{s.t.} \quad \begin{cases} -\mathbf{t} \leq \mathbf{C}\mathbf{x} \leq \mathbf{t} \\ \mathbf{B}\mathbf{x} = \mathbf{m} \end{cases},$$

where $\mathbf{1}$ denotes an "all-one" vector whose elements are all ones. It can be further turned into an LP canonical form as

$$\mathbf{x}^*, \mathbf{t}^* = \underset{\mathbf{x}, \mathbf{t}}{\operatorname{argmin}} \begin{bmatrix} \mathbf{0}^\top & \mathbf{1}^\top \end{bmatrix} \begin{bmatrix} \mathbf{x} \\ \mathbf{t} \end{bmatrix} \quad \text{s.t.} \quad \begin{cases} \begin{bmatrix} -\mathbf{C} & \mathbf{I} \\ \mathbf{C} & \mathbf{I} \end{bmatrix} \begin{bmatrix} \mathbf{x} \\ \mathbf{t} \end{bmatrix} \geq \mathbf{0} \\ \mathbf{B}\mathbf{x} = \mathbf{m} \end{cases},$$

in which $\mathbf{0}$ is an all-zero vector whose elements are all zeros. Various software packages are available for solving the canonical form of LP, such as CVXOPT.[2] Once the solution $\mathbf{x}^*$ is obtained, we have the estimates for surface normal $\tilde{\mathbf{n}}^*$ by taking the first three elements of $\mathbf{x}^*$, while $\mathbf{t}^*$ may be discarded unless it is necessary.

Unconstrained ℓ_1 Norm Minimization Eq. (5.17) via IRLS

An unconstrained ℓ_1 norm minimization problem can be efficiently solved by IRLS [12, 13]. Using the same notations $\mathbf{x}$, $\mathbf{C}$, and $\mathbf{L}$ above, Eq. (5.17) can be written as

---

[2]CVXOPT: https://cvxopt.org/.

$$\mathbf{x}^* = \underset{\mathbf{x}}{\operatorname{argmin}} \left\{ \|\mathbf{Cx}\|_1 + \lambda \|\mathbf{m} - \mathbf{Bx}\|_2^2 \right\}. \tag{5.19}$$

The basic idea of the IRLS-based method is to cast the ℓ_1 norm minimization problem into iteratively solving weighted least squares problems. In our case, the ℓ_1 norm term $\|\mathbf{Cx}\|_1$ can be written with an appropriate nonnegative diagonal weight matrix $\mathbf{W}$ as

$$\|\mathbf{Cx}\|_1 = \|\mathbf{WCx}\|_2^2 = \mathbf{x}^\top \mathbf{C}^\top \mathbf{W}^\top \mathbf{WCx}.$$

To see this, let us define a residual vector $\mathbf{r} \in \mathbb{R}^f$ as $\mathbf{r} = \mathbf{Cx}$. Also, let w_i and r_i denote the ith diagonal element of $\mathbf{W}$ and the ith element of $\mathbf{r}$, respectively. Since

$$\|\mathbf{Cx}\|_1 = \|\mathbf{r}\|_1 = \sum_i |r_i| = \sum_i |r_i|^{-1} |r_i|^2,$$

and $r_i \in \mathbb{R}$, we can rewrite this using w_i as

$$\sum_i |r_i|^{-1} |r_i|^2 = \sum_i |r_i|^{-1} r_i^2 = \sum_i w_i^2 \left(\mathbf{C}(i,:)\mathbf{x} \right) = \mathbf{x}^\top \mathbf{C}^\top \mathbf{W}^\top \mathbf{WCx},$$

where $\mathbf{C}(i,:)$ indicates the ith row vector of $\mathbf{C}$. Here, the weight element w_i is defined as

$$w_i = |r_i|^{-\frac{1}{2}}. \tag{5.20}$$

The IRLS algorithm iteratively updates the weight matrix $\mathbf{W}$ so as to satisfy the relationship in Eq. (5.20).

The IRLS method determines the weight matrix $\mathbf{W}$ over iterations. Here, we describe the sketch of the IRLS-based solution method for Eq. (5.19).

1. Initialize weight matrix $\mathbf{W}$ by an identity matrix: $\mathbf{W} \leftarrow \mathbf{I}$.
2. Solve a weighted least squares problem and update $\mathbf{x}$ as

$$\mathbf{x}_{(t+1)} \leftarrow \underset{\mathbf{x}}{\operatorname{argmin}} \left\{ \mathbf{x}^\top \mathbf{C}^\top \mathbf{W}_{(t)}^\top \mathbf{W}_{(t)} \mathbf{Cx} + \lambda \left(\mathbf{m} - \mathbf{Bx} \right)^\top \left(\mathbf{m} - \mathbf{Bx} \right) \right\},$$

where the subscript (t) indicates the tth iteration.
3. If $\mathbf{x}$ does not change much from the previous iteration, i.e., $\|\mathbf{x}_{(t+1)} - \mathbf{x}_{(t)}\|_2 < \tau$ for a small positive value τ, return $\mathbf{x}_{(t+1)}$ as the solution $\mathbf{x}^*$ and terminate.
4. Compute residual $\mathbf{r} = \mathbf{Cx}_{(t+1)}$.
5. Update weight matrix $\mathbf{W}_{(t+1)}$ by $w_i = \frac{1}{|r_i|^{\frac{1}{2}} + \epsilon}$. Here, the weight update rule Eq. (5.20) is slightly modified by adding a small positive value ϵ to avoid zero division.

As in the constrained ℓ_1 norm minimization case, the first three elements of $\mathbf{x}^*$ form the surface normal estimate $\tilde{\mathbf{n}}$. A more detailed algorithmic explanation with a general ℓ_p norm case ($1 \le p < 2$) can be found in [14].

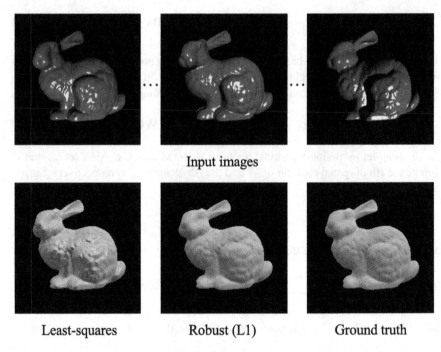

Input images

Least-squares   Robust (L1)   Ground truth

Fig. 5.5 Robust photometric stereo (ℓ_1 residual minimization) applied to a scene with strong specular reflections. While the conventional least squares method is affected by specular high-light, robust photometric stereo method effectively ignores specular high-light, yielding a result closer to the ground truth

There are advanced approximation methods for ℓ_0 norm minimization other than ℓ_1 norm approximation. One of them is a method of sparse Bayesian learning (SBL) [15, 16], which was used for solving the photometric stereo problem in the past [9, 10]. Although the relaxed form of ℓ_0 by SBL is non-convex, it is reported that the results are more accurate than convex ℓ_1 norm approximations.

Figure 5.5 shows the robust photometric stereo based on ℓ_1 residual minimization applied to a scene with strong specular reflections. The top row of the figure shows some of the input images, and in the bottom row from left to right, it shows the surface normal estimates with least squares photometric stereo and robust photometric stereo, and the ground truth surface normal. As seen in the figure, while the result of least squares photometric stereo is significantly affected by non-Lambertian specular observations, robust photometric stereo effectively ignores specular high-light to achieve faithful estimation. It shows the strength of the robust photometric stereo in handling outliers.

5.3.2 Robust Photometric Stereo via Robust Principal Component Analysis

There is another thread of work to exploit the sparsity of the corruption matrix $\mathbf{E}$. From Eq. (5.6), we know that matrix $\mathbf{LN}$ is *low rank* because both $\mathbf{L}$ and $\mathbf{N}$ are at most rank 3. Therefore, we can regard $\mathbf{LN}$ in Eq. (5.12) as a low-rank matrix $\mathbf{A}$. We can thus formulate the robust photometric stereo problem as a robust principal component analysis (R-PCA) [8, 17] problem as

$$\mathbf{A}^*, \mathbf{E}^* = \underset{\mathbf{A},\mathbf{E}}{\text{argmin}} \{\text{rank}(\mathbf{A}) + \lambda\|\mathbf{E}\|_0\} \quad \text{s.t.} \quad \mathbf{M} = \mathbf{A} + \mathbf{E}, \qquad (5.21)$$

with an assumption of sparse corruption $\mathbf{E}$. The solution for the low-rank component $\mathbf{A}^*$ represents the diffuse observations that are free from sparse outliers.

Unfortunately, the above problem is highly non-convex and discontinuous due to both rank() function and ℓ_0 norm term. However, again, we can use a convex surrogate of the problem written as

$$\mathbf{A}^*, \mathbf{E}^* = \underset{\mathbf{A},\mathbf{E}}{\text{argmin}} \{\|\mathbf{A}\|_* + \lambda\|\mathbf{E}\|_1\} \quad \text{s.t.} \quad \mathbf{M} = \mathbf{A} + \mathbf{E}, \qquad (5.22)$$

in which $\|\cdot\|_*$ indicates the nuclear norm (also known as trace norm), defined as the summation of matrix's singular values as $\|\mathbf{X}\|_* = \sum_i \sigma_i(\mathbf{X})$. Various solution techniques are known for this problem [18–20].

With the computed $\mathbf{A}^*$, we can rely on the conventional photometric stereo, such as Eq. (5.10), for solving $\mathbf{A}^* = \mathbf{LN}$ with known light matrix $\mathbf{L}$. There is a chance that rank($\mathbf{A}^*$) becomes greater than 3 because of non-Lambertian diffuse reflectances. While how to best deal with such cases has not been fully understood yet, one possible approach would be to perform rank-3 approximation to the low-rank component $\mathbf{A}^*$.

5.4 Uncalibrated Photometric Stereo

So far, we have discussed the case where the light directions $\mathbf{L}$ are calibrated. Uncalibrated photometric stereo is a problem of recovering surface normal $\mathbf{N}$ from a set of images illuminated under *unknown* light directions $\mathbf{L}$. Hayakawa showed that the shape $\mathbf{N}$ and light $\mathbf{L}$ can be obtained up to a linear ambiguity [21]. Namely, the observation matrix $\mathbf{M}$ can be factored by singular value decomposition (SVD) as

$$\mathbf{M} = \mathbf{U}\Sigma\mathbf{V}^\top = \left(\mathbf{U}\Sigma^{\frac{1}{2}}\right)\left(\Sigma^{\frac{1}{2}}\mathbf{V}^\top\right) = \mathbf{L}'\mathbf{N}', \qquad (5.23)$$

where $\mathbf{L}' = \left(\mathbf{U}\Sigma^{\frac{1}{2}}\right)$ and $\mathbf{N}' = \left(\Sigma^{\frac{1}{2}}\mathbf{V}^\top\right)$ are the light directions and surface normal. Although the SVD is unique, there is an *ambiguity* in terms of surface normal $\mathbf{N}'$

and light directions $\mathbf{L}'$. They may differ from the true light directions $\mathbf{L}$ and surface normal $\mathbf{N}$ by an arbitrary invertible 3×3 matrix $\mathbf{H}$ as

$$\mathbf{LN} = \left(\mathbf{L}'\mathbf{H}\right)\left(\mathbf{H}^{-1}\mathbf{N}'\right). \qquad (5.24)$$

This ambiguity is known as *shape-light ambiguity*, which is inherent when the lighting conditions are unknown.

It has been later pointed out that if the integrability constraint [22] is accounted over the shape $\mathbf{N}$, the general linear ambiguity can be reduced to a generalized basrelief (GBR) ambiguity [23] in the form of

$$\mathbf{H} = \begin{bmatrix} 1 & 0 & 0 \\ 0 & 1 & 0 \\ \mu & \nu & \lambda \end{bmatrix}. \qquad (5.25)$$

This 3 degrees-of-freedom (DoF) ambiguity is preferred over the general 9-DoF ambiguity if one is to disambiguate the infinitely many solutions. Figure 5.6 depicts the GBR shape-light ambiguity. In the figure, it can be seen that the same appearances can be produced by different combinations of light, albedo, and surface normal.

Light direction & intensity	Albedo	Surface normal	Sideview of the shape	Appearance

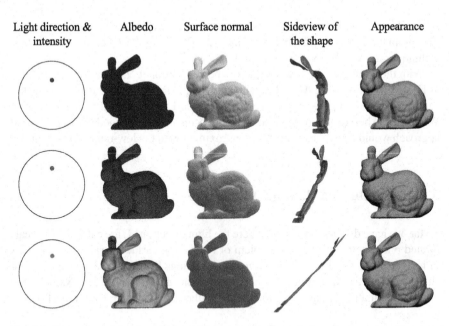

Fig. 5.6 Illustration of the shape-light ambiguity in uncalibrated photometric stereo. The same appearances can be produced by a set of different light (redder the brighter), albedo, and surface normal

5.5 Semi-calibrated Photometric Stereo

It has been recently understood that, for a Lambertian photometric stereo of Woodham [3], calibration of the light source brightnesses is actually unnecessary, but only the knowledge of light directions suffices to obtain a unique solution [24, 25]. The problem setting and the solution method are collectively called *semi-calibrated photometric stereo*.

With an abuse of notation, we begin with Eq. (5.2) but the proportionality is changed to equality by assuming that the light source brightness E_i additionally represents the global scaling. A measured pixel brightness m is then written as

$$m_{ij} = \rho_j E_i \mathbf{l}_i^\top \mathbf{n}_j, \tag{5.26}$$

where i and j are indices for the light direction and surface point (corresponding to a pixel) location. $\mathbf{l}_i \in \mathcal{S}^2$ and $\mathbf{n}_j \in \mathcal{S}^2$ are unit column vectors of the ith light direction and surface normal of the jth surface point, ρ_j is a Lambertian diffuse albedo, and E_i is the light source brightness (and sensor exposure).

In the matrix form for representing all the scene points and light directions at a time, it can be written as

$$\mathbf{M} = \mathbf{ELN}^\top \mathbf{P}. \tag{5.27}$$

The matrix $\mathbf{M} \in \mathbb{R}^{f \times p}$ is an observation matrix, where f and p are the number of images and pixels, respectively. $\mathbf{E}$ is an $f \times f$ diagonal light brightness matrix containing E_i in its elements, i.e., $\mathbf{E} = \mathrm{diag}\left(E_i, \ldots, E_f\right)$, where diag() creates a diagonal matrix from the entries. $\mathbf{L}$ is a $f \times 3$ light direction matrix, $\mathbf{N}$ is a $p \times 3$ surface normal, and $\mathbf{P}$ is a $p \times p$ diagonal diffuse albedo matrix whose diagonal elements are diffuse albedos, i.e., $\mathbf{P} = \mathrm{diag}\left(\rho_1, \ldots, \rho_p\right)$. With a matrix $\mathbf{B} = \mathbf{P}^\top \mathbf{N}$ representing the albedo-scaled surface normal matrix, the above equation is written as

$$\mathbf{M} = \mathbf{ELB}^\top. \tag{5.28}$$

The problem of semi-calibrated photometric stereo is to determine the albedo-scaled surface normal $\mathbf{B}$ and unknown light brightnesses (and exposures) $\mathbf{E}$ from observations $\mathbf{M}$ with the knowledge of light directions $\mathbf{L}$. Note that the light directions $\mathbf{L}$ only contains the directions but not the light source brightnesses as they are treated separately as $\mathbf{E}$.

With the least squares objective, the estimation problem can be written as energy minimization as

$$\mathbf{E}^*, \mathbf{B}^* = \underset{\mathbf{E}, \mathbf{B}}{\mathrm{argmin}} \, \|\mathbf{M} - \mathbf{ELB}^\top\|_F^2. \tag{5.29}$$

The inside of the squared Frobenius norm is a bilinear function of $\mathbf{E}$ and $\mathbf{B}$, which results in a non-convex optimization problem in general. However, it has been shown that by exploiting the special structure of $\mathbf{E}$, one can derive a unique solution up to a global scaling [24, 25].

Specifically, the semi-calibrated photometric stereo method [24, 25] exploits the property that the light brightness matrix $\mathbf{E}$ is always invertible as it is a positive diagonal matrix. This is because each diagonal element of $\mathbf{E}$ represents the corresponding light brightness that is always positive. Therefore, there always exists $\mathbf{E}^{-1}$. Then the original bilinear form of Eq. (5.28) can be rewritten as $\mathbf{E}^{-1}\mathbf{M} = \mathbf{L}\mathbf{B}^\top$.

Given observations $\mathbf{M}$ and light directions $\mathbf{L}$, it can therefore be viewed as a Lyapunov equation $(\mathbf{AX} + \mathbf{XB} = \mathbf{C})$ as

$$\mathbf{E}^{-1}\mathbf{M} - \mathbf{L}\mathbf{B}^\top = \mathbf{0}. \tag{5.30}$$

By vectorizing unknown variables $\mathbf{E}^{-1}$ and $\mathbf{B}^\top$, it can be written as

$$\left[\text{diag}(\mathbf{m}_1) \mid \cdots \mid \text{diag}(\mathbf{m}_p) \right]^\top \mathbf{E}^{-1}\mathbf{1} - \left(\mathbf{I}_p \otimes \mathbf{L}\right) \text{vec}\left(\mathbf{B}^\top\right) = \mathbf{0}, \tag{5.31}$$

where $\text{vec}(\cdot)$ and $\otimes$ are vectorization and Kronecker product operators, respectively. $\mathbf{1}$ indicates a vector whose elements are all one; therefore, $\mathbf{E}^{-1}\mathbf{1}$ results in a vector that consists of diagonal elements of $\mathbf{E}^{-1}$. $\mathbf{I}_p$ is a $p \times p$ identity matrix, and $\mathbf{m}_j$ is a measurement vector at the jth pixel location, i.e., the jth column vector of observation matrix $\mathbf{M}$. By concatenating matrices and vectors in Eq. (5.30), a homogeneous system of equations can be obtained as

$$\underbrace{\left[-\mathbf{I}_p \otimes \mathbf{L} \mid \left[\text{diag}(\mathbf{m}_1)\mid \cdots \mid\text{diag}(\mathbf{m}_p)\right]^\top\right]}_{\mathbf{D}} \underbrace{\begin{bmatrix} \text{vec}(\mathbf{B}^\top) \\ \mathbf{E}^{-1}\mathbf{1} \end{bmatrix}}_{\mathbf{y}} = \mathbf{0}, \tag{5.32}$$

where $\mathbf{D} \in \mathbb{R}^{pf \times (3p+f)}$ is a known sparse matrix, and $\mathbf{y} \in \mathbb{R}^{(3p+f) \times 1}$ is a vector of unknowns that we wish to find. The homogeneous system always has a trivial solution $\mathbf{y} = \mathbf{0}$. To attain a unique nontrivial solution up to scale, the matrix $\mathbf{D}$ should have a one-dimensional null space, i.e., $\text{rank}(\mathbf{D}) = 3p + f - 1$. Under the condition, we can obtain the solution $\mathbf{y}$ with a constraint $\|\mathbf{y}\|_2^2 = 1$ via SVD. Namely, with the SVD factorization $\mathbf{D} = \mathbf{U}\boldsymbol{\Sigma}\mathbf{V}^\top$, the singular value matrix $\boldsymbol{\Sigma}$ will have exactly one diagonal entry being zero, and the corresponding column vector of $\mathbf{V}$ becomes the solution $\mathbf{y}^*$.

There is a necessary condition for the semi-calibrated photometric stereo problem to have a meaningful solution. As stated, when the rank of matrix $\mathbf{D} \in \mathbb{R}^{pf \times (3p+f)}$ in Eq. (5.32) is $3p + f - 1$, a unique solution up to scale can be obtained via SVD. In this case, the row dimension pf should be equal to or greater than the rank. Thus, the minimum condition to have a unique solution up to scale is $pf \geq (3p + f - 1)$, which can be rewritten as

$$(p - 1)(f - 3) \geq 2. \tag{5.33}$$

The condition indicates that semi-calibrated photometric stereo requires more than one pixel that correspond to different surface normals (i.e., $p > 1$) and more than three distinct light directions (i.e., $f > 3$). In other words, even if we do not have

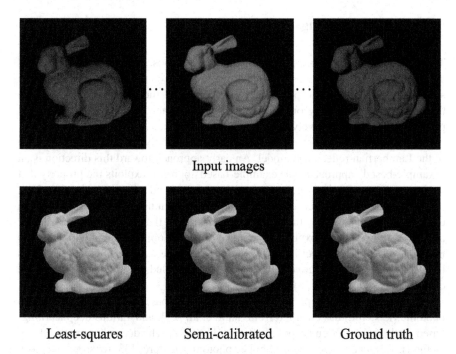

Input images

Least-squares   Semi-calibrated   Ground truth

Fig. 5.7 Semi-calibrated photometric stereo applied to a scene observed under varying light brightnesses (or, equivalently, varying exposure times). While the conventional least squares method produce rather flattened surface, the semi-calibrated photometric stereo method is uninfluenced by the variation of the brightness, yielding a result closer to the ground truth

an access to light source brightnesses, the photometric stereo problem can be solved with 4 different light directions and 2 pixels that correspond to different surface normals via the semi-calibrated photometric stereo method.

Figure 5.7 shows an example of semi-calibrated photometric stereo applied to a scene observed under varying light brightnesses (or, equivalently, varying exposure times). The top row of the figure shows some of the input images. As we can see in the input images, the image brightnesses are different due to the different strengths of incoming light. When the light strengths are assumed uniform, or light strengths are inaccurately calibrated, the result of conventional least squares photometric stereo is affected by the unmodelled light strength fluctuations. Semi-calibrated photometric stereo, on the other hand, can faithfully estimate both the surface normal and relative strengths of incoming light across images, resulting in a surface normal map closer to the ground truth.

5.6 Further Readings

This chapter described basic photometric stereo methods, namely, Lambertian least squares photometric stereo, robust photometric stereo, uncalibrated photometric stereo, and semi-calibrated photometric stereo. All of these methods discussed in this chapter basically rely on the Lambertian reflectance model, although robust photometric stereo effectively treats deviations from the Lambertian model.

There are photometric stereo approaches that explicitly alleviate the assumption of the Lambertian reflectance model. An early approach toward this direction is an "example-based" approach. The example-based approach exploits the property that distinct surface points having the same surface normal and reflectance should exhibit the same appearance under a distant light for determining the surface normal. An early work along this direction is found in Horn and Ikeuchi [26] in 1984. Later, the idea is revisited with explicitly placing a reference object painted with the same material to the target object in [27]. More recently, the use of synthetic renderings has been shown to be effective and is successfully applied in a similar manner to the example-based approach [28].

In a different thread, a photometric stereo method based on deep learning (deep photometric stereo) has emerged from Santo et al. [29], and various deep learning-based methods have been proposed afterward [30–32]. The deep photometric stereo method is further extended to uncalibrated photometric stereo [33] for simultaneously determining light directions and surface normals.

References

1. Horn BK (1970) Shape from shading: a method for obtaining the shape of a smooth opaque object from one view. Technical report
2. Ikeuchi K, Horn BK (1981) Numerical shape from shading and occluding boundaries. Artif Intell 17(1–3):141–184
3. Woodham RJ (1980) Photometric method for determining surface orientation from multiple images. Opt Eng 19(1):191139
4. Ikeuchi K (1981) Determining surface orientations of specular surfaces by using the photometric stereo method. IEEE Trans Pattern Anal Mach Intell (PAMI) 6:661–669
5. Barsky S, Petrou M (2003) The 4-source photometric stereo technique for three-dimensional surfaces in the presence of highlights and shadows. IEEE Trans Pattern Anal Mach Intell (PAMI) 25(10):1239–1252
6. Mukaigawa Y, Ishii Y, Shakunaga T (2007) Analysis of photometric factors based on photometric linearization. J Opt Soc Am 24(10):3326–3334
7. Miyazaki D, Hara K, Ikeuchi K (2010) Median photometric stereo as applied to the segonko tumulus and museum objects. Int J Comput Vis (IJCV) 86(2–3):229
8. Wu L, Ganesh A, Shi B, Matsushita Y, Wang Y, Ma Y (2010) Robust photometric stereo via low-rank matrix completion and recovery. In: Proceedings of the Asian conference on computer vision (ACCV). Springer, pp 703–717
9. Ikehata S, Wipf D, Matsushita Y, Aizawa K (2012) Robust photometric stereo using sparse regression. In: Proceedings of the IEEE conference on computer vision and pattern recognition (CVPR). IEEE, pp 318–325

10. Ikehata S, Wipf D, Matsushita Y, Aizawa K (2014) Photometric stereo using sparse bayesian regression for general diffuse surfaces. IEEE Trans Pattern Anal Mach Intell (PAMI) 36(9):1816–1831
11. Gentle JE (2007) Matrix algebra: theory, computations, and applications in statistics, 1st edn. Springer Publishing Company, Incorporated
12. Lawson CL (1961) Contribution to the theory of linear least maximum approximation. PhD thesis, PhD dissertation, UCLA
13. Gorodnitsky IF, Rao BD (1997) Sparse signal reconstruction from limited data using focuss: a re-weighted minimum norm algorithm. IEEE Trans Image Process (TIP) 45(3):600–616
14. Samejima M, Matsushita Y (2016) Fast general norm approximation via iteratively reweighted least squares. In: Proceedings of the Asian conference on computer vision (ACCV) workshops. Springer, pp 207–221
15. Tipping ME (2001) Sparse Bayesian learning and the relevance vector machine. J Mach Learn Res 1(Jun):211–244
16. Wipf DP, Palmer J, Rao BD (2004) Perspectives on sparse Bayesian learning. In: Advances in neural information processing systems, pp 249–256
17. Candès EJ, Li X, Ma Y, Wright J (2011) Robust principal component analysis? J ACM (JACM) 58(3):1–37
18. Lin Z, Chen M, Ma Y (2010) The augmented lagrange multiplier method for exact recovery of corrupted low-rank matrices. arXiv:1009.5055
19. Ji S, Ye J (2009) An accelerated gradient method for trace norm minimization. In: Proceedings of the international conference on machine learning. ACM, pp 457–464
20. Oh TH, Matsushita Y, Tai YW, Kweon IS (2017) Fast randomized singular value thresholding for low-rank optimization. IEEE Trans Pattern Anal Mach Intell (PAMI) 40(2):376–391
21. Hayakawa H (1994) Photometric stereo under a light source with arbitrary motion. J Opt Soc Am A 11. https://doi.org/10.1364/JOSAA.11.003079
22. Horn BK, Brooks MJ (1986) The variational approach to shape from shading. Comput Vis Graph Image Process 33(2):174–208
23. Belhumeur PN, Kriegman DJ, Yuille AL (1999) The bas-relief ambiguity. Int J Comput Vis (IJCV) 35(1):33–44
24. Cho D, Matsushita Y, Tai YW, Kweon I (2016) Photometric stereo under non-uniform light intensities and exposures. In: Proceedings of the European conference on computer vision (ECCV). Springer, pp 170–186
25. Cho D, Matsushita Y, Tai YW, Kweon IS (2018) Semi-calibrated photometric stereo. IEEE Trans Pattern Anal Mach Intell (PAMI) 42(1):232–245
26. Horn BK, Ikeuchi K (1984) The mechanical manipulation of randomly oriented parts. Sci Am 251(2):100–113
27. Hertzmann A, Seitz SM (2005) Example-based photometric stereo: shape reconstruction with general, varying brdfs. IEEE Trans Pattern Anal Mach Intell (PAMI) 27(8):1254–1264
28. Hui Z, Sankaranarayanan AC (2016) Shape and spatially-varying reflectance estimation from virtual exemplars. IEEE Trans Pattern Anal Mach Intell (PAMI) 39(10):2060–2073
29. Santo H, Samejima M, Sugano Y, Shi B, Matsushita Y (2017) Deep photometric stereo network. In: Proceedings of the international conference on computer vision (ICCV) workshops, pp 501–509
30. Chen G, Han K, Wong KYK (2018) Ps-fcn: A flexible learning framework for photometric stereo. In: Proceedings of the European conference on computer vision (ECCV), pp 3–18
31. Ikehata S (2018) Cnn-ps: Cnn-based photometric stereo for general non-convex surfaces. In: Proceedings of the European conference on computer vision (ECCV), pp 3–18
32. Taniai T, Maehara T (2018) Neural inverse rendering for general reflectance photometric stereo. In: Proceedings of the international conference on machine learning, pp 4864–4873
33. Chen G, Han K, Shi B, Matsushita Y, Wong KYK (2019) Self-calibrating deep photometric stereo networks. In: Proceedings of the IEEE conference on computer vision and pattern recognition (CVPR), pp 8739–8747

Chapter 6
Structured Light

The structured light method is a very popular way to find correspondences between coordinates of a camera image and those of a projector image. This method class uses spatial or temporal image patterns as clues to obtain correspondence information. The basic approaches are described in Sect. 2.5, and various extensions have been proposed to apply structured light methods to various situations. In this chapter, the methods are classified based on projection patterns, and several extension methods are introduced.

6.1 Triangulation-Based Laser Sensors

Today, many types of 3D sensors are available in the market. A significant portion of those sensors rely on triangulation between active light and imaging sensors. The approach is often called *active stereo*, and we call sensors based on it *active-stereo sensors*. Figure 6.1 shows major examples.

The first example (Fig. 6.1a) is a point-laser sensor, where a laser spot projected by a point laser is captured by a camera. Triangulation can be performed with the projected laser-light ray and the captured ray. Because this sensor can capture a single point of a scene for one image capture, the point laser is often moved with respect to two axes to capture certain areas.

The second example (Fig. 6.1b) is a line-laser sensor, which projects curves onto a scene via a line laser to be observed by a camera. 3D points on the projected curves are obtained via geometrical reasoning (e.g., ray tracing) to estimate the intersection of a ray and a plane. In this configuration, the line laser is often rotated with respect to the camera to capture areas of a scene. Another major setup uses a laser and a camera fixed on a rig; the rig is moved with respect to the scene.

The third example (Fig. 6.1c) is a pattern-projection sensor, where a 2D pattern projected on a scene from a 2D structured light source (e.g., a video projector) is observed by a camera. This type can be thought of as a stereo–camera pair, where

© Springer Nature Switzerland AG 2020
K. Ikeuchi et al., *Active Lighting and Its Application for Computer Vision*,
Advances in Computer Vision and Pattern Recognition,
https://doi.org/10.1007/978-3-030-56577-0_6

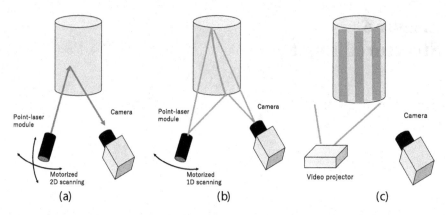

Fig. 6.1 Active-stereo sensors

one camera of a normal pair is replaced by a projector. In this configuration, the correspondence problem, in which each observed pattern feature captured by the camera is associated with a certain position of the projected 2D pattern, is the core of 3D reconstruction methods. Because this sensor can capture areas of a scene, the projector and the camera are normally fixed to each other, such as with the Kinect v1 or the Intel RealSense [1].

In this section, active-stereo sensors using point lasers and those using line lasers are explained. Active-stereo sensors that use 2D pattern projectors are explained later, because there are many techniques related to this type to address the correspondence problem.

6.1.1   Triangulation with Point Lasers (0D)

For measuring a 3D point using a calibrated point laser and a camera, the laser beam is first projected onto the target. Then, the laser spot is captured by the camera. Because the laser light has a peaky wavelength distribution, users can obtain a better signal-to-noise ratio (SNR) by mounting a narrowband filter onto the camera lens. The laser spot on the captured image becomes a small region with high brightness values, and the region can be detected via simple image processing (e.g., thresholding). To obtain better 3D position quality, the center of gravity of the image spot is often calculated to localize the laser spot in sub-pixel accuracies:

$$\mathbf{p}_s = \frac{\sum_{\mathbf{p} \in \mathcal{R}} I(\mathbf{p})}{\sum_{\mathbf{p} \in \mathcal{R}} (I(\mathbf{p}) \, \mathbf{p})} \tag{6.1}$$

where $\mathcal{R}$ is the set of pixels of the detected region, $\mathbf{p}_s$ is the sub-pixel location of the laser spot, and $I()$ is the function from a pixel to the brightness value. There

Fig. 6.2 Triangulation of a point laser and a camera

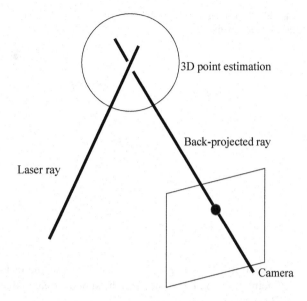

3D point estimation

Back-projected ray

Laser ray

Camera

are several other methods of sub-pixel localization of a laser spot, such as fitting the brightness distribution to Gaussian function or parabola function.

When the laser spot on an image is localized, the 2D position is back-projected using camera parameters to obtain the observed ray in 3D space. Similarly, the projected laser ray should also be identified as a 3D line, using the calibration result of the point laser.

Triangulation with the configuration of a point laser and a pixel position of a camera is a similar setup that uses triangulation with camera-to-camera stereo, where two ray lines are used to estimate a 3D crossing point. Unfortunately, with noise, those two rays may not cross (Fig. 6.2). This problem has been intensively researched in computer vision research [2–7].

A well-known triangulation method is the *midpoint method*. For the two nonparallel rays, there is a line that crosses each perpendicularly. The midpoint method determines the 3D point to the midpoint of the two intersections on this line. See Fig. 6.3.

Calculation of this point is simple. Following the notations of Yan et al. [7], the camera ray is represented by the camera center, $\mathbf{o}_1$, and the ray direction vector, $\mathbf{b}_1$. The laser ray is represented by the light source, $\mathbf{o}_2$, and the light direction, $\mathbf{b}_2$. Then, the midpoint, $\mathbf{p}$, can be calculated by minimizing

$$e(\mathbf{p}) = \sum_{i=1}^{2} \|(\mathbf{I} - \mathbf{b}_i \mathbf{b}_i^\top)(\mathbf{p} - \mathbf{o}_i)\|, \tag{6.2}$$

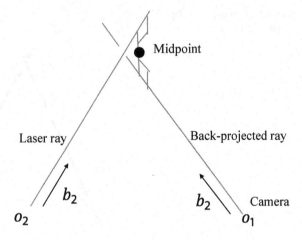

Fig. 6.3 Triangulation using the midpoint method

because $\mathbf{b}_i\mathbf{b}_i^\top$ projects $\mathbf{p}$ to each of the rays of the orthogonal projection, and the midpoint of two points minimizes the sum of squared distances from the two points.

By defining $\mathbf{B}_i = \mathbf{I} - \mathbf{b}_i\mathbf{b}_1^\top$,

$$e(\mathbf{p}) = \sum_{i=1}^{2} \|\mathbf{B}_i(\mathbf{p} - \mathbf{o}_i)\|. \tag{6.3}$$

Minimization of $e(\mathbf{p})$ can be achieved by solving equations where the gradient of $e(\mathbf{p})$ equals the zero vector. Thus,

$$(\mathbf{B}_1+\mathbf{B}_2)\mathbf{p} = \mathbf{B}_1\mathbf{o}_1+\mathbf{B}_2\mathbf{o}_2. \tag{6.4}$$

Because $(\mathbf{B}_1+\mathbf{B}_2)$ is a 3×3 matrix, and the right side is a constant vector, this equation can be solved analytically.

Another well-known approach is based on the minimization of reprojection errors, which are used in the bundle-adjustment approach. Because this approach assumes that each ray is associated with a camera, we replace the point-laser projector with a camera by treating the direction of the laser light as the observed pixel point, as shown in Fig. 6.4. Via this approach, the estimated 3D point is assumed to be projected onto each of the cameras, and the differences between these projected points and the observed 2D points are reprojection errors. The 3D point can be estimated by minimizing the sum of the squared distance errors on these image points.

For a two-view configuration, Lindstrom [6] proposed an efficient iterative algorithm. According to their notations, $\mathbf{S}$ is defined as

$$\mathbf{S} \equiv \begin{bmatrix} 1 & 0 & 0 \\ 0 & 1 & 0 \end{bmatrix}, \tag{6.5}$$

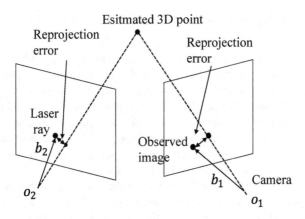

Fig. 6.4 Triangulation with reprojection error minimization

1: **procedure** LINDSTROMITERATIVE($\mathbf{x}_0, \mathbf{x}'_0, \mathbf{E}$)   ▷ Iterative method of [171]
2:     **for** $k = 1, 2, \cdots$ **do**
3:         $\mathbf{n}_k \leftarrow \mathbf{SEx}'_{k-1}$
4:         $\mathbf{n}'_k \leftarrow \mathbf{SE}^\top \mathbf{x}'_{k-1}$
5:         $a_k \leftarrow \mathbf{n}_k^\top \tilde{\mathbf{E}} \mathbf{n}'_k$
6:         $b_k \leftarrow \frac{1}{2}\{\mathbf{n}_1^\top \mathbf{n}_k + (\mathbf{n}'_1)^\top \mathbf{n}'_k\}$
7:         $c_k \leftarrow \mathbf{x}_0^\top \mathbf{E} \mathbf{x}'_0$
8:         $d_k \leftarrow \sqrt{b_k^2 - a_k c_k}$
9:         $\lambda_k \leftarrow \frac{c_k}{b_k + \mathrm{sgn}(b_k) d_k}$
10:        $\Delta \mathbf{x}_k \leftarrow \lambda_k \mathbf{n}_k$
11:        $\Delta \mathbf{x}'_k \leftarrow \lambda_k \mathbf{n}'_k$
12:        $\mathbf{x}_k \leftarrow \mathbf{x}_0 - \mathbf{S}^\top \Delta \mathbf{x}_k$
13:        $\mathbf{x}'_k \leftarrow \mathbf{x}'_0 - \mathbf{S}^\top \Delta \mathbf{x}'_k$

Fig. 6.5 Iterative triangulation method of Lindstrom [6]

and $\tilde{\mathbf{E}}$ is defined as

$$\tilde{\mathbf{E}} \equiv \mathbf{SES}^\top \tag{6.6}$$

which is the upper left 2×2 sub-matrix of the essential matrix, $\mathbf{E}$. The algorithm is shown in Fig. 6.5, where the input $\mathbf{x}_0$ is the 2D image position in a normalized camera represented in homogenous coordinates, $\mathbf{x}'_0$ is the direction of the laser represented in a 2D normalized camera image in homogenous coordinates, and $\mathbf{E}$ is the essential matrix. The outputs after k iterations are corrected 2D coordinates $\mathbf{x}_k$ and $\mathbf{x}'_k$. After the solution converges, the midpoint algorithm can be applied for $\mathbf{x}_k$ and $\mathbf{x}'_k$ to obtain the optimum 3D point. The algorithm of [6] converges with two iterations when the two cameras are set parallel.

6.1.2   Triangulation with Line Lasers (1D)

A major category of the triangulation-based laser sensor is its combination of a line laser and a camera, as shown in Sect. 6.1. In this configuration, the positions of the 3D points can be calculated as an intersection of a back-projected ray and a laser plane. Thus, we do not need to solve the correspondence problem, and we can get dense depth images by rotating the laser projector around a single axis. Because it can be a good compromise between the difficulty of the correspondence problem and the number of points acquired for a single image capture, many commercial 3D laser scanners have been developed with this configuration.

To measure 3D points with the configuration, the line-laser beam is projected onto the target, as shown in Fig. 6.6. The illuminated points form 3D curves on the scene surface, and the 3D curve is captured by the camera as 2D curves.

As with the case of laser-spot detection, we can detect the region of the laser curve via thresholding. It is detected as narrow regions having certain widths. Localizing the captured 2D curves is more difficult than localizing a laser spot. Thus, we must obtain a skeletonized curve from the detected regions. Through this process, the direction along the 2D curve and other directions are treated differently.

A good strategy of skeletonization of the curves involves scanning the image along its epipolar lines, as shown in Fig. 6.7. The epipolar lines can be considered as directions of the set of the laser beams that form the lights of the line laser on the

Fig. 6.6 Triangulation with line lasers

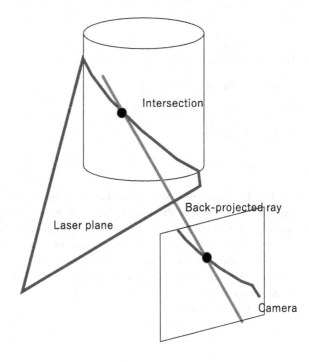

Fig. 6.7 Laser-curve localization

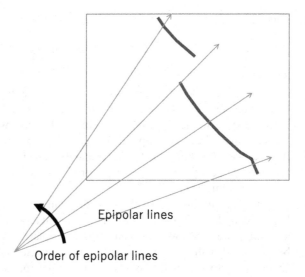

camera 2D image. Thus, an epipolar line does not become parallel to the detected laser curves, and it does not cross the laser curves more than once.

Along the epipolar lines, the center of the laser intensity profile can be sampled as a 1D function of the laser profile. The center of the laser profile can be calculated by estimating the center of gravity of intensity or by fitting the profile to Gaussian or parabola functions. The curve can be represented as line segments, via an ordered sampling that is associated with ordered epipolar lines. Their order is shown in Fig. 6.7.

When the 2D curve on the image is localized, triangulation calculation between the 2D curves and the line laser is far simpler than with the case of point lasers. The 3D positions can be calculated as intersections of the laser plane and the back-projected rays of the 2D curves. During this process, each point on the detected 2D curves is back-projected using the camera parameter. The back-projected ray can be represented as

$$\mathbf{x} = \mathbf{o} + t\mathbf{b}, \tag{6.7}$$

where $\mathbf{o}$ is the camera position, $\mathbf{b}$ is the direction vector, and t is a parameter with $t > 0$. The calibrated laser plane can be represented as

$$\mathbf{a}^\top \mathbf{x} = 1 \tag{6.8}$$

where $\mathbf{a}$ is the parameter vector of the plane. The calculation of the intersection the plane and the line can be solved by

$$t = \frac{1 - \mathbf{a}^\top \mathbf{o}}{\mathbf{a}^\top \mathbf{b}}. \tag{6.9}$$

Thus,

$$\mathbf{x} = \mathbf{o} + \frac{1 - \mathbf{a}^\top \mathbf{o}}{\mathbf{a}^\top \mathbf{b}} \mathbf{b}. \qquad (6.10)$$

As described, the line laser must be moved to capture the whole scene observed by the camera. One option is fixing the line laser and the camera into a rig and moving the rig to the scene. An example of the moving rig is a car-mounted road scanner, where the road bumps are measured from within the car driving on the road. Another option is fixing the line laser and the camera and moving the target object. For measurement purposes of factory products, items on a belt conveyer are often measured in this setup.

Note that, different from the case of point lasers, the line-laser configuration has no redundancy in the calculation of triangulation. In other words, even if there are calibration errors with the camera or laser-plane parameters, the 3D points can be calculated with no contradictions. This contrasts the point-laser case, because calibration errors of point-laser systems can be found as non-crossing rays. The large distance between the back-projected ray and the laser beam suggests large calibration errors. However, in the case of line-laser configuration, calibration errors result in distorted output shapes. Thus, users must remain cognizant of whether accurate calibrations have been performed.

6.1.3   Self-calibrating Laser Scanners

Triangulation-based sensors having point or line lasers should also have a method (e.g., servo motors) of rotating the laser devices to scan entire visible surfaces from the camera. This can imply large costs, because the system must use a servo-controlled motor with a synchronized image-capture process, and the laser ray or the laser planes should be perfectly calibrated including the model of the controlled motion of the laser devices.

To alleviate this problem, some researchers have proposed using self-calibrating laser devices. For example, in the configuration of Fig. 6.8, the laser device is rotated without camera synchronization. Instead of modeling and pre-calibrating the controlled laser projector, one or more reference planes are pre-calibrated and used to estimate the position of the laser planes online. In the case of Fig. 6.8, if the positions of the two reference planes are known, each observed point on the reference plane can be localized in 3D space using the same mathematical calculation as that of the light-section method. If three noncollinear points are observed, the laser plane can be estimated. In this setup, the laser device can even be moved by hand.

Chu et al. [8] proposed another solution (Fig. 6.9). They added a cubic frame to the field of view and placed the object inside the frame. They then emitted a line beam to the object, detecting a bright point on the cube to estimate the position of the beam plane and the 3D information.

Fig. 6.8 Self-calibrating
laser scanner

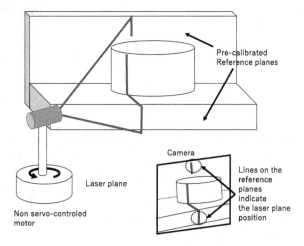

Fig. 6.9 Self-calibrating
laser scanner [8]

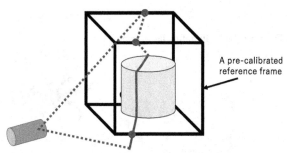

Takatsuka et al. [9] adopted a more active method (Fig. 6.10). They put LED markers on the sensor and captured it using a single camera to estimate the sensor position. They used a laser pointer as a sensor and proposed a clever method of efficient 3D data calculation using the detected LED markers.

Furukawa and Kawasaki placed LED markers on the sensor and captured the markers using a single camera to estimate the laser plane [10], as shown in Fig. 6.11. In this method, the devices required for this system were a video camera and a laser projector mounted with LED markers for pose detection. The laser projector in our system emits a laser plane rather than a laser beam.

In their method, the rotational pose of the line-laser projector was first estimated via vanishing point estimation from the detected marker positions. Then, the pose of the projector is refined by minimizing the reprojection errors of the marker positions. When the pose of the line-laser projector was estimated, the 3D reconstruction could be performed via the light-section method. The results are shown in Fig. 6.12.

Self-Calibrating Laser Scanners Without Reference Objects

Furukawa et al.proposed methods for self-calibrating a line-laser scanner without reference objects [11].

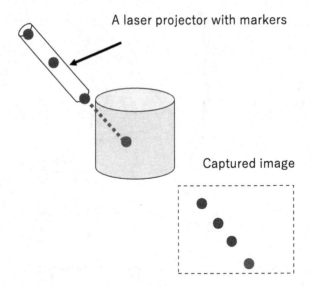

Fig. 6.10 Self-calibrating laser scanner [9]

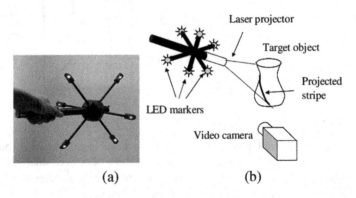

Fig. 6.11 Self-calibrating line-laser scanner using markers [10]

Geometrically, a line laser is a device for extracting a set of points existing on the same plane (i.e., coplanar points). By utilizing the coplanarities extracted by line lasers, the laser planes can be self-calibrated online. Figure 6.13a shows an example of the system used to observe coplanarities, wherein the target scene is captured using a fixed camera while the scene is lit by a line laser. Using this system, many coplanarities can be observed as curves that are reflections of the laser on the scene by capturing sequences of images while moving the laser projector. In the work of Furukawa et al., they also used planes in the scene for other types of coplanarities.

By capturing multiple frames while accumulating the curves of multiple images, a number of intersection points were extracted, as shown in Fig. 6.13b. These intersection points can be used for constraints between different laser or scene planes.

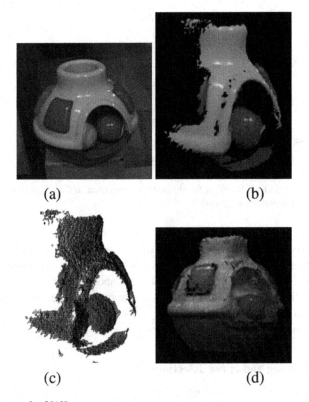

Fig. 6.12 Scan result of [10]

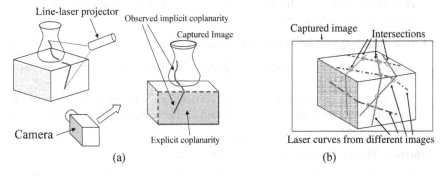

Fig. 6.13 Self-calibration system using coplanarities. **a** Observation of implicit and explicit coplanarities. **b** Points of intersection of the laser curves

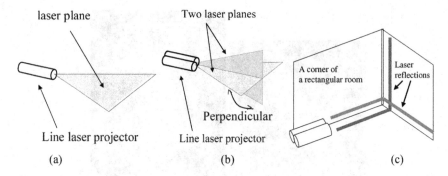

Fig. 6.14 Laser-projecting device. **a** Single-laser configuration. **b** Crosshair-laser configuration. **c** Adjusting crosshair-laser configuration

They showed that one intersection point of the scene could be interpreted as a linear equation between two planes (i.e., laser planes or scene planes). They also showed that, generally, the accumulated system of these above linear equations has 4-DoF solutions. To fix the 4-DoF ambiguity, they proposed the use of a perpendicularly mounted set of line lasers, as shown in Fig. 6.14.

Scanning the 3D scene using the system is accomplished via the following processes:

- **Image capturing and curve detection**

  – A sequence of images is captured using the fixed camera while moving the line laser back and forth manually. Meanwhile, laser curves are extracted, and information of those extracted curves is stored instead of the captured images to reduce data storage size.

- **Curve sampling and intersection detection**

  – From all captured laser curves, a smaller number of curves are sampled to reduce computational costs.
  – By aggregating the sampled laser curves on a common image plane, intersection points between those curves can be obtained. These intersection points exist on multiple planes.

- **Solving 3D information up to 4-DoF ambiguity**

  – By constructing a set of linear equations and solving them, as shown in [11], the laser planes of the selected curves are reconstructed up to 4-DoF indeterminacies from the coordinates of the intersection points on the image plane.

- **Determining 4-DoF ambiguity**

  – Metric reconstruction methods with metric constraints are applied to eliminate 4-DoF indeterminacies. Such information is given by two ways:

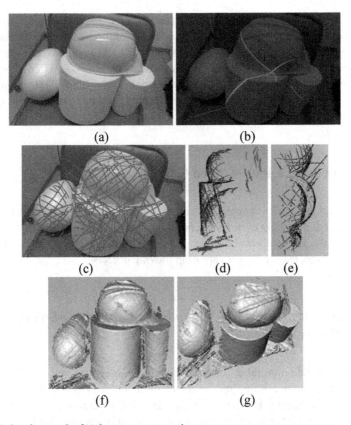

Fig. 6.15 Another result of real scene reconstruction

For the single-laser configuration, the information can be manually retrieved from the scene. This includes selecting regions of explicit coplanarity, as shown in Fig. 6.13a.

For the crosshair-laser configuration, the information is automatically given from the relationship between two lasers (i.e., the angle of 90°).

- **3D reconstruction using light-section method**

  – Using the reconstructed laser-plane parameters and the extracted laser curves, 3D shapes of the selected laser curves are reconstructed using a light-section method, as explained in Sect. 6.1.2.

Furukawa et al. used a laser-projecting device that emitted two laser planes orthogonal to each other (i.e., crosshair-laser configuration). They selected 23 images and reconstructed the 3D shape. They also conducted a dense reconstruction. Image processing for detecting laser curves and their intersection points were the same as in their previous experiment. Figure 6.15a–g shows all inputs and results. They used 43 laser curves with 975 intersection points for projective reconstruction. Three lines were removed, because the intersection points on each were collinear. They used 20 orthogonalities for metric reconstruction, and 213,437 points were reconstructed using the dense reconstruction process. The processing time was 0.078 s for projective reconstruction and 0.016 s for metric reconstruction. 0.37 s were required for dense reconstruction for 2,160 frames. It can be seen from this that an arbitrary shape was successfully reconstructed.

6.2   Structured Light of Area (2D)

Structured light systems use a camera that captures 2D images. If a 2D Structured light pattern is used, many 2D correspondences distributed in an image can be obtained simultaneously. The methods introduced in this section use 2D patterns to make use of this advantage, especially for capturing moving objects.

6.2.1   Temporal Encoding Methods for Capturing Moving Objects

Methods based on temporal encoding use multiple images to find correspondence, meaning they require a long time for acquisition, and they poorly capture moving objects. To overcome the issue, several approaches have been proposed to reduce the acquisition time. One reduces the number of required images needed to be captured and switching patterns at high speed.

Encoding by Stripe Boundary

The method proposed by Hall-Holt and Rusinkiewicz [12, 13] determines the correspondence between camera and projector by projecting a black and white stripe pattern and assigning a code to the boundary of stripes. As shown in Fig. 6.16a, four images were taken of different stripe patterns and binarized. A 4-bit code as obtained for each pixel. If the code of adjacent pixels was different, the boundary between stripes and the 3D plane defined by the boundary could be uniquely determined. Depending on the codes of adjacent pixels, the boundary may not be detected from the camera image. As a result, it is possible to use 110 types of patterns from a 4 + 4-bit code using codes on both sides of the boundary. Thus, the method classified 110 steps of distance by decoding the codes. This was combined with spatial encoding, because decoding was not completed inside each pixel, and the information around the boundary was required. Thus, a large number of distance steps were encoded with as few as four images.

In the experimental system, the pattern was changed at 30 or 60 Hz using a DLP video projector, and the camera captured 60 FPS in synchronization with the projector. To obtain a 4-bit code from an image sequence, the positions of the boundaries

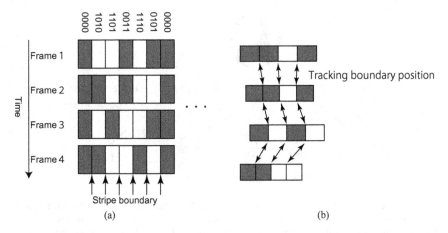

Fig. 6.16 Encoding by the code on the stripe boundary. **a** Projection of four stripe patterns and assigned codes to boundaries. **b** When the observation target moves during acquisition, matching between frames is necessary, because the position of the boundary changes

were fixed during the four frames. However, when the observation target was moving, they were not fixed on the camera image. Therefore, as shown in Fig. 6.16b, tracking of boundaries between frames was performed to realize the observation of a moving object. Because a simple method of selecting the closest boundary was used for matching in the proposed method, the speed of the movement of the observation target was limited and did not cope with the fast movement.

Fast Phase Shifting

This is a method of high-speed phase shifting that projects a fringe pattern in which the brightness change is a sine wave. Weise et al. [14] developed a system that modified a DLP projector and changed the pattern at high speed. A DLP projector has a color wheel, and a color image is generated by time-dividing the projected light into four colors, RGB+white. When the color wheel is removed, the system projects different black and white patterns that are time-divided. When an image of a different RGB pattern is input from a personal computer (PC), the projector automatically switches the pattern. There is no need for a PC to control the switching patterns. In the experimental system, the pattern is changed at 120 Hz, and the synchronized camera captures images to realize high-speed pattern imaging. Additionally, as shown in Fig. 6.17, an additional camera is used to simultaneously capture the texture while projecting a pattern. Because the texture camera captures images by taking a long exposure time, the images are captured without a pattern by canceling the projected pattern, which has a high frequency in time series.

To perform phase unwrapping, a method was proposed that removes the ambiguity of periodic coordinates via stereo vision using two cameras. As a result, the correspondence can be uniquely determined without additional images. In the proposed method, an image was acquired three times by shifting the phase by $(2/3)\pi$,

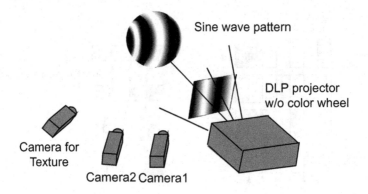

Fig. 6.17 Setup in the method of Weise et al. [14]. The phase of an image is determined by projecting a sine-wave pattern from a projector and capturing it three times at different phases. The cycle ambiguity is removed by using two cameras. Additionally, simultaneous measurement of textures is realized by using a camera for capturing texture with a long exposure time

but the phase difference did not become $(2/3)\pi$ when the observation target was moving. If the phase was calculated as it was, a plane is measured as a wavy shape. Therefore, considering movement, the phase was corrected so that the phase difference was $(2/3)\pi$. The proposed method linearizes the problem by assuming that the shape is locally planar and that the motion is a constant-velocity translation. It estimates the phase using the least squares method. If this assumption is not met, or if the motion is fast, and the linearization approximation does not work well, the phase correction will not be accurate. This method was based on phase shift, which is a temporal encoding method, and can measure the distance for each pixel of a camera image and generate a high-density-shape model. To overcome the difficulty of measuring moving objects via temporal encoding, this method introduces phase estimation correction by assuming linearity with nearby points.

Combination of Active Lighting and Passive Stereo

In the active methods, one of two cameras in a stereo vision system is replaced with a light source. There are active-stereo methods that use active light sources [15–17]. As mentioned, if there is no texture on the target object, it is difficult to search for corresponding points using passive stereo methods. To tackle this issue, there are active methods that use an active light source to project a texture onto the target object to facilitate the corresponding point search in the passive stereo method [15]. The light source's texture changes over time instead of requiring an alternative camera in this approach. It has the advantage of not requiring light-source calibration.

To reliably search for corresponding points using the passive stereo method, the correspondence of the texture pattern of each point must be unique. To obtain the uniqueness of the texture pattern, both spatial and temporal patterns can be used. If a temporal pattern is used, the spatial window area used for stereo matching can be reduced to improve the spatial resolution [15, 16]. However, because the temporal resolution is degraded, it becomes difficult to reconstruct fast-moving objects.

Additionally, there are several issues with this approach, such as the projected pattern requiring synchronization with stereo cameras at high speed and an expensive spatiotemporal image matching process.

Random Pattern Projection by DMD

While the display cycle of a DLP projector is about 120 Hz, the DMD element used to generate the pattern can be switched at tens of microseconds to change the pattern much faster. Narasimhan et al. [18] used a high-speed camera to capture a high-speed pattern generated by DMD and used a brightness value obtained from multiple images as a feature vector to match the projected pattern shown in Fig. 6.18a.

In the proposed method, a brightness value of 0–255 is input to the projector, and a high-speed camera capable of imaging 300–3000 FPS is used. The brightness of each frame changes according to turning the DMD on and off, as shown in Fig. 6.18b. The change pattern is unique to the brightness value of 0–255, as shown in Fig. 6.18c. Therefore, a feature vector is formed by the brightness values of multiple frames, and the correspondence between the brightness value and the observed brightness can be determined uniquely. Because commercially available DLP projectors cannot finely control DMD operation, the projected pattern differs depending on the projector. Therefore, the method examines the pattern corresponding to the brightness value in advance. The image input to the projector uses a stripe pattern with random colors. Because dark colors are difficult to distinguish from camera images, the number of colors that can be used is about 160.

Basically, synchronization between the camera and the projector is required for matching. However, the method proposed placing a known object in the field of view to remove the need to explicitly synchronize, because matching is performed based on that. The proposed method uses 20 images for matching, and we can assume that the observation target does not move in the frame for matching, because the image is taken using a high-speed camera, and motion compensation is not required. However, there is an upper limit to the motion speed that can be measured. For example, when 20 images taken at 1,000 FPS are used, the exposure time is 20 ms, and the movement must be sufficiently small within that time. Additionally, it is essential to use a high-speed camera at a frame rate of about 300 FPS. Therefore, it is not suitable for observation of slow movement.

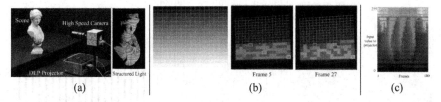

Fig. 6.18 Active shape-measurement system [18]. **a** System with high-speed pattern switching using DMD. **b** When an input image (left) is given to the projector, the brightness captured with a high-speed camera changes because of DMD pattern switching. **c** Relationship between the brightness value and brightness of the input image over time

6.2.2 One-Shot Methods

The previous section introduced a method of correct measurement that switched patterns at high speeds and compensated for them by assuming simple motions. However, in practice, there is a limit to measurement by pattern switching, because it does not well-handle very fast or simple movements. Therefore, a one-shot measurement method using only one input image has attracted attention as a method that can cope with any movement. This is a technique for projecting a fixed pattern and restoring the shape from a single image taken by a camera. If the one-shot method is used, it is theoretically possible to measure the shape corresponding to the fast movement by increasing the shutter speed.

In this section, typical one-shot measurement methods are introduced by classifying them into three types, and application methods to moving object measurement are explained.

Line-Based Method

In this approach, a line pattern is projected, and the correspondence is determined using the intersection of a line pattern and an epipolar line. The information is embedded in the line to obtain the correspondence between the line on the projection pattern and the detected line. An example of this method is shown in Fig. 2.22a.

From the methods proposed thus far, those that directly embed information into a line are often taken. The simplest method uses the brightness value as identification (ID) as is (i.e., direct method). Since this method uses brightness values directly, only a few examples under special conditions have been proposed, because it is considered difficult to put to practical use because of its low accuracy and reliability. Thus, to use it for practical measurement of moving objects, it is common to increase the stability by using the brightness values before and after the epipolar line at the same time. For example, there is a method that uses brightness values for each RGB [21] (Fig. 6.19a) and a method that embeds information into colored lines using a De Bruijn sequence [22]. However, the method of giving information to the brightness value itself is generally low and unstable, because the brightness value of the pattern is directly affected by the texture of the object surface to be projected. Therefore,

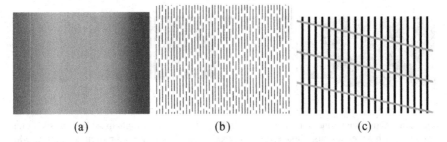

(a)   (b)   (c)

Fig. 6.19 Examples of patterns of line-based methods: **a** Direct methods. **b** Dotted pattern from [19]. **c** Koninckx's method [20]

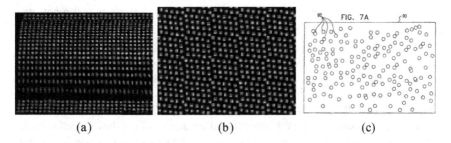

(a)   (b)   (c)

Fig. 6.20 Examples of patterns of area-based methods. **a** M-array from [26]. **b** Pattern by 2 × 2 Cell from [27]. **c** Pattern of Kinect v1 [28]

a method using only the binary on/off value of the pattern can be considered to avoid this problem. For example, there is a method [23] that embeds information as a broken-line pattern by embedding a line (Fig. 6.19b), or a method [24] that embeds information in variable line thicknesses. However, Koninckx et al. proposed a method that realized dense-shape reconstruction from a single image using a simple stripe pattern [20] (Fig. 6.19c). This method was based on local shape restoration by extracting dense striped repetition of vertical line patterns combined with global position estimations using auxiliary sparse horizontal-line patterns for reconstruction. The number, position, and inclination of the horizontal lines were adjusted so that there was only one intersection of the vertical and horizontal lines on each epipolar line. In other words, with this method, line-ID information is embedded only at the intersections of sparse horizontal lines and vertical lines, and when detection of horizontal-line patterns fails or discontinuities occur in the vertical stripe patterns, the shape is not reconstructed. Consequently, objects that can be measured are limited to those having smooth shapes and few texture patterns. Additionally, a method that considers temporal continuity [25] has been proposed for the measurement of moving objects.

These methods are generally performed using the light-section method because of the nature of line methods. The light-section method is often used for 3D reconstruction using line lasers, etc. The 3D point is calculated as the intersection point of the viewing ray of the camera image and the plane formed by the projecting of a line in the projector image.

Area-Based Methods

With line-based methods, it is assumed that correspondence information is embedded along the lines or that decoding is performed on a line-by-line basis. Thus, a line of a certain length must be detected, or a continuous pattern string must be observed to some extent. Because this is a big limitation in practice, continuous brightness information is often used with binary information about the existence of the line. However, it is easily affected by texture and becomes unstable. Methods for solving this problem have been proposed by embedding information in a 2D area around each point [26, 27, 29–32]. Examples of this approach are shown in Fig. 2.22b. The correspondence is obtained by matching the 2D features.

Morano et al. proposed a method using an M-array pattern that ensured uniqueness in an area of $M \times M$ pixels [26] (Fig. 6.20a). Kimura et al. proposed a method to embed corresponding-point information in a 2–2 cell using RGB [27] (Fig. 6.20b). Kinect [28] uses a pattern that maintains uniqueness within the parallax range by arranging dots well, although it is a single color (Fig. 6.20c).

Because epipolar geometry is established in area- and line-based methods, the uniqueness of the local area pattern only needs to be realized on each epipolar line. The area-based method can efficiently embed information in 2D and is relatively robust against large changes in the shape and texture of the target. However, the pattern is greatly distorted when the baseline between the camera and the projector is large to improve accuracy, or when the orientation of the surface of the target object is far from fronto-parallel. Simple area-matching methods do not work in such cases, and there are limitations, even for methods that can cope with large distortions. Thus, pattern encoding and decoding has been proposed.

Grid-Graph-Based Methods

As mentioned, in the area-based method, accuracy actually decreases because of the distortion of the pattern caused by the baseline and shape change and the influence of the object texture. However, in the case of a line pattern, as long as the lines are connected continuously, they are hardly affected by distortion. Nevertheless, the detection of long continuous lines or patterns required by the line-based method remains difficult, because the lines are inevitably cut off, owing to the influence of texture and shape discontinuities.

If the lines are connected in a graph, as shown in Fig. 6.21, the connectivity on the graph can be used according to the continuity of each line. For example, it is possible to obtain a very robust correspondence that is not easily affected by pattern distortion. Based on this idea, methods to determine the correspondence that use not only the position of the intersections between lines, but also the connections between the intersections [33–35], have been proposed. In these methods, no information is embedded in the line or area itself. Thus, the pattern is very simple, the detection is robust, and it is suitable for measuring moving objects whose shapes change frequently. An example of this method is shown in Fig. 6.22.

In the first method, vertical and horizontal lines are emitted from a single projector in the form of a grid, and the 3D shape is reconstructed based on the coplanarity condition established at the intersection of these patterns [33]. The coplanarity condition is a geometric constraint that exists at the intersections of planes on the image. It is known that the solution space has 4 DoF [36]. Because the method must find vertical and horizontal lines independently, the easiest way to distinguish them is by using different colors for vertical and horizontal lines, as shown in Fig. 6.22a.

To increase the stability of the solution, methods, such as those using the De Bruijn sequence for color [34] as shown in Fig. 6.22b and pattern interval [35], have been proposed. A De Bruijn sequence is a sequence of q^n with a number of symbols, q, and the length, n, needed to uniquely identify the position. A De Bruijn sequence

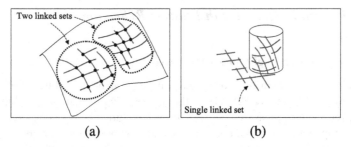

Fig. 6.21 Examples of connected graphs. **a** Two sets of linked sets. **b** Single set of linked sets

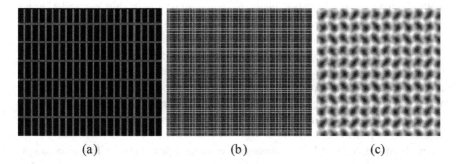

Fig. 6.22 Examples of a pattern of grid-based methods. **a** Using different colors for vertical/horizontal lines. **b** A pattern based on De Bruijn sequence. **c** A pattern with wave lines of single color

of length n and number of symbols, q, is a sequence of length q^n, and, if a partial sequence of length n is observed, the sequence has the characteristics from which the position in the sequence can be uniquely identified. This characteristic is used to identify color codes for active scans [22, 29, 31].

Color use is very powerful for obtaining pattern information from an acquired image. However, determining colors can be erroneous, because color is affected by the characteristics of the light source, object, and camera. Thus, it is important to match the pattern of a single color. Thus, a wave-grid pattern [37] has been proposed, as shown in Fig. 6.22c. The wave line is a sinusoidal pattern that is periodic and self-recurring. The grid of wave lines can contain information for finding correspondences. The proposed method uses the intersection points of vertical and horizontal wave lines as feature points. If the interval of the vertical wave lines is not equal to the integral multiple of the horizontal wavelength, the intersection points appear at different phases on the wave pattern. This means that the local pattern around the intersection point has a local uniqueness and that it can be used as a discriminative feature.

6.2.3 Pattern Detection with Indirect Reflection

Problems of Indirect Light on 3D Measurements

When active illumination is used for shape measurement, the influence of indirect components can be a major obstacle. In the case of using structured light, the signals required for distance measurement should only be direct components, because the indirect components are noisy. For example, when a point in the scene is surrounded by surfaces having large reflectivity, the point is often considerably lit by indirect illuminations, causing measurement errors. In particular, with shape-measurement methods that directly use pixel-intensity values (e.g., phase-shift methods), noise caused by indirect components becomes a big problem.

An example for which the influence of the indirect light component is significant is a scene in which a diffuse reflecting plane is combined with an angle that is concave from the camera position. In this situation, light irradiated from the vicinity of the camera is reflected multiple times, providing specular reflections near the concave angle. Then, it reaches the camera as indirect components. Because this is caused by specular reflections, the indirect components can be strong, greatly affecting the 3D measurement processes.

Furthermore, indirect light components can cause problems for ToF-camera systems, for which the travel time of light emitted from the camera and reflected at target surfaces is measured for each pixel. However, if light is reflected in the scene multiple times, thus traveling longer than the direct light components, it can cause disturbances of measurement values.

3D Measurement Methods Stable Against Indirect Components

As described, indirect components of light-transport functions are of low frequency with respect to projector pixels. This indicates that high-frequency changes in the projector pattern do not affect the brightness of the observed indirect components. Thus, if multiple patterns having the same low-frequency distributions and different high-frequency distributions are projected, the indirect components of the observed images can be observed as constants (e.g., bias signals) for each pixel.

In the phase-shift method, the bias component of observed multiple sinusoidal patterns for each pixel can be estimated and canceled using properties of sine signals. Using this process, Gupta et al. proposed "micro phase shift method" [38], in which only sinusoidal patterns having high spatial frequencies are projected, and the results are more robust against effects of indirect illuminations than those of a naive phase-shift method.

A set of Gray code patterns is often used as a binary code pattern set for estimating correspondences between a projector and a camera. However, naive Gray code patterns include images having various (from low to high) spatial frequencies (Fig. 2.19). In such a pattern set, the low-frequency patterns can be strongly influenced by the indirect light transport, causing code-reading errors. On the other hand, Gray code patterns may include excessive high-frequency patterns, causing the problem where the high-frequency pattern may be blurred out in the observed pattern because of defocused projectors or cameras.

To address this problem, patterns having special frequencies appropriate for Gray code projections are used, and they are decoded in the presence of indirect light and defocused [39–41]. Using such patterns, the effects of indirect light transports and the defocusing of cameras and projectors can be reduced.

There is another method of projecting an arbitrary pattern other than binary code, but it occurs at the cost of more projections. When n patterns are projected, the luminance value at each pixel can be regarded as an n-dimensional feature vector. Because each pixel in the projection pattern and in the observed image sequence are, respectively, represented by such n-dimensional feature vectors, the correspondence between the projection pattern and the observed image can be estimated by matching these feature vectors. Such a method has a high DoF in pattern design. Using this technique, Couture et al. proposed a projection pattern that was less affected by indirect illumination, obtained via spatial frequency analysis [42].

Generally, indirect components of images have a lower frequency spatial component than to those of direct components. Thus, it is possible to estimate direct/indirect separation from a single image by using a priori knowledge. Art et al. proposed a method to separate an input image into direct and indirect components by decomposing the image into base images of various frequencies [43].

6.2.4 Extension of 3D Reconstruction by Pattern Projection

3D Reconstruction for Fast Moving Objects

High-speed pattern-switching methods have attracted great attention and have been studied as active 3D measurement methods for moving objects. However, because most methods assume that motion can be linearly approximated in a short time, only very slow motion can be measured for nonrigid objects (e.g., humans and liquids). Recently, many practical measurement methods using the one-shot method have been proposed. Thus, high-speed shape measurement has been realized. For example, Sagawa et al. [34] used a grid-based one-shot scan technique to measure the shape of a bursting balloon captured at 1,000 FPS, as shown in Fig. 6.23. In [44], they improved the method to generate a dense reconstruction of water splashes and human-face deformations, as shown in Fig. 6.24. The method using a wave-grid pattern [37] also generated dense 3D reconstructions of a moving person, as shown in Fig. 6.25.

Reconstruction of Whole 3D Shape

With 3D shape measurements that use images from one viewpoint, only a part of the 3D shape can be measured. However, for many applications, measuring the entire 3D shape is of great importance. Thus, it is necessary to simultaneously measure 3D shapes from multiple viewpoints. With measurements that use an active light

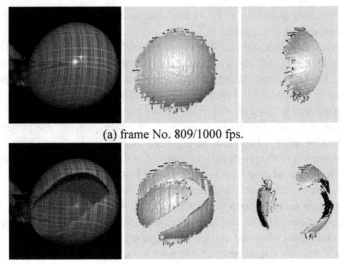

(a) frame No. 809/1000 fps.

(b) frame No. 824/1000 fps.

Fig. 6.23 Results of a bursting balloon

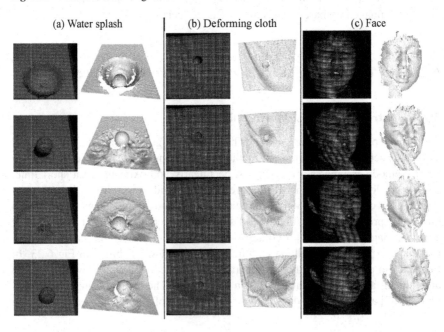

(a) Water splash   (b) Deforming cloth   (c) Face

Fig. 6.24 Dense 3D reconstruction. **a** Water splash. **b** Ball and cloth. **c** Human face

Fig. 6.25 The 3D shapes of a punching scene are reconstructed. The detailed shape of the cloth is successfully obtained

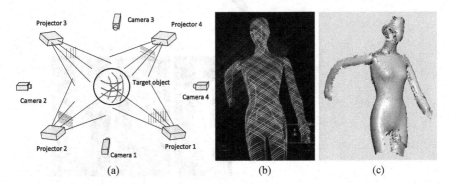

Fig. 6.26 System for capturing an entire 3D shape [45]. **a** System overview. **b** Input image. **c** Reconstruction result

source, it is necessary to project multiple light sources to the target object to perform omnidirectional measurement. In this case, because multiple light sources interfere, it is necessary to separate them. To do this, a method based on time division or one based on colors should be considered. However, there are problems, such as the small number of available colors and complicated devices. Furukawa et al. [45] proposed a method of measuring the entire shape of a dynamic object by arranging projectors and cameras alternately in a circle (Fig. 6.26). At this time, because the projected light sources are simple parallel patterns, it is easy to separate them and detect them. The reconstruction is performed in a similar way as that of grid reconstruction [33–35], which uses the intersection of line patterns captured by each camera. Considering the connectivity between patterns at intersections, the 3D position of a connected pattern set has nearly 1D ambiguity. By resolving this ambiguity by collation with multiple cameras, a stable 3D reconstruction was realized. Figure 6.27 shows the restoration

Fig. 6.27 3D reconstruction
of the entire shape of a
dynamic scene

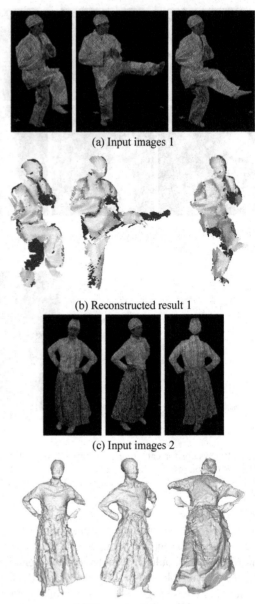

(a) Input images 1

(b) Reconstructed result 1

(c) Input images 2

(d) Reconstructed result 2

(a)   (b)   (c)

Fig. 6.28 DOE-based pattern projector. **a** Experimental pattern projector developed by combining a 532 nm laser light source and the designed DOE. **b** Viewing angle of a generated pattern of ∼ 40°. **c** Zoom-in of the pattern

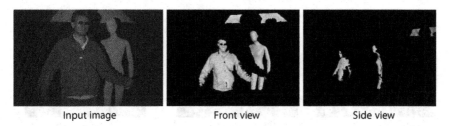

Input image   Front view   Side view

Fig. 6.29 An example of 3D reconstruction using a laser pattern projector

result. It can be seen that the entire shape was reconstructed, despite the complicated patterns overlapping. Additional constraints for cases of multiple projectors have been proposed to improve the robustness of matching [46–48].

Structured Light with Various Projectors

An easy way to generate a structured light pattern is to use an off-the-shelf video projector, which can change pattern controlling by using a PC. A typical LCD projector has a lamp bulb as the light source and masks to generate patterns. Because more than 70% pixels are blocked by the mask, the large part of the light energy is discarded, and the power efficiency is low. Moreover, the size of a projector can be too large for practical uses, such as mounting on a robot's head. An option to tackle the problems of light efficiency and projector size is using laser light, which can contribute to improvements in power-efficiency and projector size. Instead of generating patterns via masking, DOEs can be used. One works as a beam splitter that divides an input beam into a large number of output beams, which generally have about 80% efficiency of input light power. A DOE can be designed to generate various patterns, and Fig. 6.28 shows an experimental pattern projector [49] that generates wave-grid patterns for one-shot 3D reconstruction. A pattern is represented by a set of dots, as shown in Fig. 6.28, having flat brightness and low distortion. Figure 6.29 shows an example of 3D reconstruction using a laser projector.

Another challenge is the reconstruction of a small-object 3D shape. A stereo microscope observes a target from two viewpoints, corresponding to eyepieces. Many

Fig. 6.30 Microscope with pattern projection

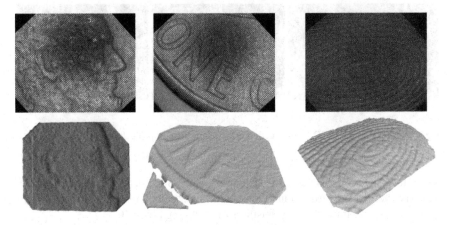

Fig. 6.31 3D reconstruction using a microscope with pattern projection

stereo microscopes can equip a camera by utilizing the position of one eyepiece to record the images. Additionally, the eyepiece can equip a reticle (i.e., a glass plate with visual marks to provide measurement). If light enters from one eyepiece having a reticle with a mask pattern, a camera can observe the pattern as structured light from the other perspective. If a one-shot pattern is printed on the reticle, the microscope can be used as a projector–camera system without additional modification. Figure 6.30 shows an experimental system that projects the wave-grid pattern and observes it with a camera. Figure 6.31 shows the results of the 3D reconstruction of a coin with a fingerprint. The system can observe 50 μm object bumps.

References

1. Keselman L, Iselin Woodfill J, Grunnet-Jepsen A, Bhowmik A (2017) Intel realsense stereo-scopic depth cameras. In: Proceedings of the IEEE conference on computer vision and pattern recognition (CVPR) workshops, pp 1–10
2. Kanazawa Y, Kanatani K (1995) Reliability of 3-D reconstruction by stereo vision. IEICE Trans Inf Syst 78(10):1301–1306
3. Hartley RI, Sturm P (1997) Triangulation. Comput Vis Image Underst J (CVIU) 68(2):146–157
4. Hartley R, Zisserman A (2003) Multiple view geometry in computer vision. Cambridge University Press
5. Kanatani K, Sugaya Y, Niitsuma H (2008) Triangulation from two views revisited: hartley-sturm versus optimal correction. Practice 4:5
6. Lindstrom P (2010) Triangulation made easy. In: Proceedings of the IEEE conference on computer vision and pattern recognition (CVPR). IEEE, pp 1554–1561
7. Yang K, Fang W, Zhao Y, Deng N (2019) Iteratively reweighted midpoint method for fast multiple view triangulation. IEEE Robot Autom Lett 4(2):708–715
8. Chu CW, Hwang S, Jung SK (2001) Calibration-free approach to 3D reconstruction using light stripe projections on a cube frame. In: Proceedings third international conference on 3-D digital imaging and modeling. IEEE, pp 13–19
9. Takatsuka M, West GA, Venkatesh S, Caelli TM (1999) Low-cost interactive active monocular range finder. In: Proceedings of the IEEE conference on computer vision and pattern recognition (CVPR), vol 1. IEEE, pp 444–449
10. Furukawa R, Kawasaki H (2003) Interactive shape acquisition using marker attached laser projector. In: Proceedings of the fourth international conference on 3-D digital imaging and modeling, 2003. 3DIM 2003. IEEE, pp 491–498
11. Furukawa R, Kawasaki H (2009) Laser range scanner based on self-calibration techniques using coplanarities and metric constraints. Comput Vis Image Underst J (CVIU) 113(11):1118–1129
12. Hall-Holt O, Rusinkiewicz S (2001) Stripe boundary codes for real-time structured-light range scanning of moving objects. In: Proceedings of the international conference on computer vision (ICCV), vol 2. IEEE, pp 359–366
13. Rusinkiewicz S, Hall-Holt O, Levoy M (2002) Real-time 3D model acquisition. ACM Trans Graph (TOG) 21(3):438–446
14. Weise T, Leibe B, Van Gool L (2007) Fast 3D scanning with automatic motion compensation. In: Proceedings of the IEEE conference on computer vision and pattern recognition (CVPR). IEEE, pp 1–8
15. Zhang L, Snavely N, Curless B, Seitz SM (2004) Spacetime faces: high resolution capture for modeling and animation. In: Proceedings of SIGGRAPH, pp 548–558
16. DAVIS J (2005) Spacetime stereo: a unifying framework for depth from triangulation. IEEE Trans Pattern Anal Mach Intell (PAMI) 27(2):296–302
17. Young M, Beeson E, Davis J, Rusinkiewicz S, Ramamoorthi R (2007) Viewpoint-coded structured light. In: Proceedings of the IEEE conference on computer vision and pattern recognition (CVPR). IEEE, pp 1–8
18. Narasimhan SG, Koppal SJ, Yamazaki S (2008) Temporal dithering of illumination for fast active vision. In: Proceedings of the European conference on computer vision (ECCV). Springer, pp 830–844
19. Batlle J, Mouaddib E, Salvi J (1998) Recent progress in coded structured light as a technique to solve the correspondence problem: a survey. Pattern Recogn 31(7):963–982
20. Koninckx TP, Geys I, Jaeggli T, Van Gool L (2004) A graph cut based adaptive structured light approach for real-time range acquisition. In: Proceedings of the 2nd international symposium on 3D data processing, visualization and transmission, 2004. 3DPVT 2004. IEEE, pp 413–421
21. Tajima J, Iwakawa M (1990) 3-D data acquisition by rainbow range finder. In: Proceedings of the international conference on pattern recognition (ICPR), vol 1. IEEE, pp 309–313

22. Zhang L, Curless B, Seitz SM (2002) Rapid shape acquisition using color structured light and multi-pass dynamic programming. In: Proceedings of the first international symposium on 3D data processing visualization and transmission. IEEE, pp 24–36
23. Maruyama M, Abe S (1993) Range sensing by projecting multiple slits with random cuts. IEEE Trans Pattern Anal Mach Intell (PAMI) 15(6):647–651
24. Artec (2007) United States Patent Application 2009005924
25. Koninckx TP, Van Gool L (2006) Real-time range acquisition by adaptive structured light. IEEE Trans Pattern Anal Mach Intell (PAMI) 28(3):432–445
26. Morano RA, Ozturk C, Conn R, Dubin S, Zietz S, Nissanov J (1998) Structured light using pseudorandom codes. IEEE Trans Pattern Anal Mach Intell (PAMI) 20(3):322–327
27. Kimura M, Mochimaru M, Kanade T (2008) Measurement of 3D foot shape deformation in motion. In: Proceedings of the 5th ACM/IEEE international workshop on projector camera systems, pp 1–8
28. Primesense (2010) United States Patent Application US 2010/0118123
29. Je C, Lee SW, Park RH (2004) High-contrast color-stripe pattern for rapid structured-light range imaging. In: Proceedings of the European conference on computer vision (ECCV). Springer, pp 95–107
30. Pan J, Huang PS, Chiang FP (2005) Color-coded binary fringe projection technique for 3-D shape measurement. Opt Eng 44(2):023606
31. Salvi J, Batlle J, Mouaddib E (1998) A robust-coded pattern projection for dynamic 3D scene measurement. Pattern Recogn Lett 19(11):1055–1065
32. Vuylsteke P, Oosterlinck A (1990) Range image acquisition with a single binary-encoded light pattern. IEEE Trans Pattern Anal Mach Intell (PAMI) 12(2):148–164
33. Kawasaki H, Furukawa R, Sagawa R, Yagi Y (2008) Dynamic scene shape reconstruction using a single structured light pattern. In: Proceedings of the IEEE conference on computer vision and pattern recognition (CVPR). IEEE, pp 1–8
34. Sagawa R, Ota Y, Yagi Y, Furukawa R, Asada N, Kawasaki H (2009) Dense 3D reconstruction method using a single pattern for fast moving object. In: Proceedings of the international conference on computer vision (ICCV). IEEE, pp 1779–1786
35. Ulusoy AO, Calakli F, Taubin G (2009) One-shot scanning using de Bruijn spaced grids. In: Proceedings of the international conference on computer vision (ICCV) workshops. IEEE, pp 1786–1792
36. Furukawa R, Kawasaki H (2006) Self-calibration of multiple laser planes for 3D scene reconstruction. In: Third international symposium on 3D data processing, visualization, and transmission (3DPVT'06). IEEE, pp 200–207
37. Sagawa R, Sakashita K, Kasuya N, Kawasaki H, Furukawa R, Yagi Y (2012) Grid-based active stereo with single-colored wave pattern for dense one-shot 3D scan. In: 2012 second international conference on 3D imaging, modeling, processing, visualization & transmission. IEEE, pp 363–370
38. Gupta M, Nayar SK (2012) Micro phase shifting. In: Proceedings of the IEEE conference on computer vision and pattern recognition (CVPR). IEEE, pp 813–820
39. Kim D, Ryu M, Lee S (2008) Antipodal gray codes for structured light. In: Proceedings of the IEEE international conference on robotics and automation (ICRA). IEEE, pp 3016–3021
40. Gupta M, Agrawal A, Veeraraghavan A, Narasimhan SG (2011) Structured light 3d scanning in the presence of global illumination. In: Proceedings of the IEEE conference on computer vision and pattern recognition (CVPR). IEEE, pp 713–720
41. Gupta M, Agrawal A, Veeraraghavan A, Narasimhan SG (2013) A practical approach to 3D scanning in the presence of interreflections, subsurface scattering and defocus. Int J Comput Vis (IJCV) 102(1–3):33–55
42. Couture V, Martin N, Roy S (2014) Unstructured light scanning robust to indirect illumination and depth discontinuities. Int J Comput Vis (IJCV) 108(3):204–221
43. Subpa-asa A, Fu Y, Zheng Y, Amano T, Sato I (2018) Separating the direct and global components of a single image. J Inf Process 26:755–767. https://doi.org/10.2197/ipsjjip.26.755

44. Sagawa R, Kawasaki H, Kiyota S, Furukawa R (2011) Dense one-shot 3D reconstruction by detecting continuous regions with parallel line projection. In: Proceedings of the international conference on computer vision (ICCV). IEEE, pp 1911–1918

45. Furukawa R, Sagawa R, Kawasaki H, Sakashita K, Yagi Y, Asada N (2010) One-shot entire shape acquisition method using multiple projectors and cameras. In: Pacific-Rim symposium on image and video technology (PSVIT). IEEE, pp 107–114

46. Kasuya N, Sagawa R, Kawasaki H, Furukawa R (2013) Robust and accurate one-shot 3D reconstruction by 2c1p system with wave grid pattern. In: 2013 International Conference on 3D Vision-3DV 2013. IEEE, pp 247–254

47. Kasuya N, Sagawa R, Furukawa R, Kawasaki H (2013) One-shot entire shape scanning by utilizing multiple projector-camera constraints of grid patterns. In: Proceedings of the international conference on computer vision (ICCV) workshops, pp 299–306

48. Sagawa R, Kasuya N, Oki Y, Kawasaki H, Matsumoto Y, Furukawa R (2014) 4D capture using visibility information of multiple projector camera system. In: 2014 2nd international conference on 3D vision, vol 2. IEEE, pp 14–21

49. Sagawa R, Kawamura T, Furukawa R, Kawasaki H, Matsumoto Y (2014) One-shot 3D reconstruction of moving objects by projecting wave grid pattern with diffractive optical element. In: Proceedings of the 11th IMEKO symposium laser metrology for precision measurement and inspection in industry (2014)

Chapter 7
Other Shape Reconstruction Techniques

In the previous chapters, the photometric stereo and structured light methods were introduced. Although these two methods have been well studied in the computer vision field, other optical phenomena are used to provide clues for shape reconstruction. In this chapter, shape reconstruction techniques that utilize special optical phenomena (e.g., polarization, fluorescence, cast shadow, and projector defocus) are introduced.

7.1 Shape from Polarization

As mentioned in Sect. 1.1.4, polarization provides rich information about a scene. In particular, it is useful for estimating the surface normal.

A smooth surface normal can be obtained using a photometric approach. Polarization [1–3] is a commonly used characteristic of light that can be exploited for this purpose.

Unpolarized light, specularly reflected at an object surface, becomes partially polarized. The angle where $I_{\min}$ is observed represents the angle of the plane that includes the surface normal. Regarding diffuse reflection, the angle where $I_{\max}$ is observed includes the surface normal. The angle of diffuse reflection is 90° rotated from the angle of specular reflection. If we know that the object surface is either specular-dominant or diffuse-dominant, we can obtain one constraint for the surface normal from light. In other words, we seek to know whether an object surface produces a specular reflection stronger than its diffuse reflection or a diffuse reflection stronger than its specular reflection (Fig. 7.1).

If we shine unpolarized light onto an object surface, the specularly reflected light will be partially polarized. We can obtain the azimuth angle of the object surface from the angle of linear polarization when $I_{\min}$ is observed. Most people assume that diffuse reflections are not polarized. However, in reality they are slightly polarized.

© Springer Nature Switzerland AG 2020
K. Ikeuchi et al., *Active Lighting and Its Application for Computer Vision*,
Advances in Computer Vision and Pattern Recognition,
https://doi.org/10.1007/978-3-030-56577-0_7

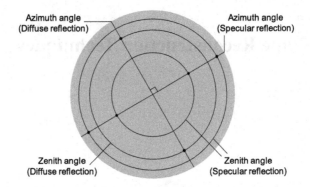

Fig. 7.1 Ambiguity of the surface normal in Gaussian sphere representation

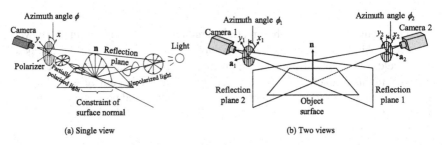

(a) Single view   (b) Two views

Fig. 7.2 Relationship between the surface normal and the reflection plane. **a** Single viewpoint. **b** Two viewpoints

The azimuth angle of a surface normal can be obtained from the angle of a linear polarizer where I_{max} is observed. Surface normal cannot be uniquely determined from a single view. Wolff and Boult [4] showed that a surface normal could be uniquely determined if we were to use an additional view (Fig. 7.2). Thus, a surface shape was obtained when observed from two views [5–8]. However, the polarization images of the two views should be analyzed at the same point on the object's surface. Rahmann and Canterakis [7] first set the initial value of corresponding points and estimated the shape, iterating this process until the shape of the object was fully estimated. Rahmann [8] also proved that the quadric surface could be obtained if the corresponding points were automatically searched.

The azimuth angle obtained from polarization has 180° ambiguity, because the linear polarizer has 180° ambiguity. Atkinson and Hancock [9] solved the ambiguity problem using polarization from two views. They locally estimated the surface shape for each candidate, and determined the corresponding point whether the candidate shape coincided between two local shapes of two views.

Wolff and Boult [4] showed that a unique surface normal could not be determined from a polarized image observed from a single viewpoint (Fig. 7.2a). Thus, Miyazaki et al. [10] observed the object from multiple viewpoints (Fig. 7.2b). To analyze the polarization state of the reflected light at the corresponding points when

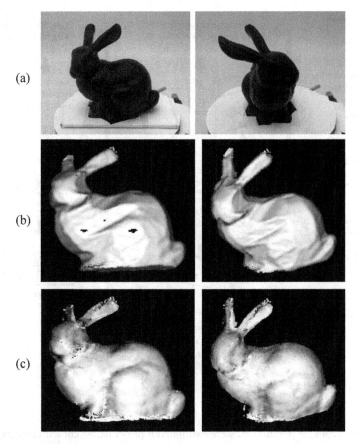

Fig. 7.3 Shape from polarization and space carving. **a** Target object. **b** Shape calculated from space carving. **c** Shape calculated from polarization and space carving

observed from multiple viewpoints, they predetermined the abstract shape using a space carving technique. Unlike a conventional photometric stereo or multi-view stereo, which cannot estimate the shape of a black specular object, their method estimated the surface normal and 3D coordinates of black specular objects using polarization analysis and space carving (Fig. 7.3).

This method [10] was not suitable for concave objects, because it depends on space carving. Miyazaki et al. [11] modified this method in order to estimate the surface normal of concave objects. The target object had a specular surface without diffuse reflection. Factories in industrial fields have a high demand for estimating the shapes of cracks, because it is quite important for quality control. Therefore, there is a great demand for estimating the shapes of concave objects of highly specular surfaces. However, it is a challenging task. Miyazaki et al. [11] estimated the surface normal of a black specular object having a concave shape by analyzing the polarization state of the reflected light in which the target object was observed from multiple

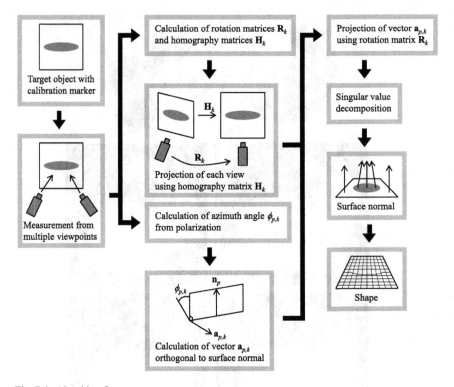

Fig. 7.4 Algorithm flow

views. The camera parameters were estimated a priori using known corresponding points, because it was assumed that the target object was almost planar. Under this assumption, the surface normal of the object was uniquely determined. Figure 7.4 shows the algorithm flow. The result of Fig. 7.5 is shown in Figs. 7.6 and 7.7. The result of Fig. 7.8 is shown in Fig. 7.9.

The depth does not change along the orientation that is orthogonal to the azimuth angle of the surface normal. The constraint using this property is called an *isodepth* constraint (Fig. 7.10) and can be proven using the definition of the azimuth angle of the differential surface with knowledge of differential geometry. Furthermore, the orientation of the azimuth angle is the steepest orientation of the depth. This constraint is represented as follows.

$$\frac{\sin \phi(x, y)}{\cos \phi(x, y)} = \frac{d(x, y+1) - d(x, y)}{d(x+1, y) - d(x, y)}. \tag{7.1}$$

Or,

$$\tan \phi = \frac{\nabla_y d}{\nabla_x d}. \tag{7.2}$$

Fig. 7.5 Target object
(ellipsoid)

Fig. 7.6 Estimated shape
(ellipsoid)

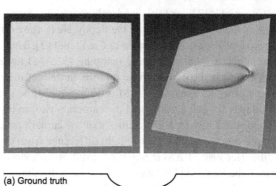

Fig. 7.7 Intersection shape
(ellipsoid). **a** Ground truth. **b**
Proposed method

(a) Ground truth

(b) Proposed method

Fig. 7.8 Target object
(stripe)

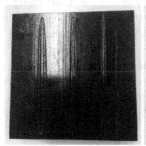

Fig. 7.9 Estimated shape
(stripe)

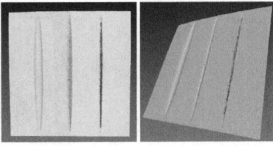

Fig. 7.10 Isodepth
constraint represented by the
azimuth angle

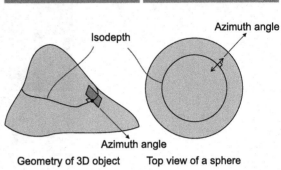

Geometry of 3D object   Top view of a sphere

Cui et al. [12] used this isodepth constraint. First, structure from motion (SfM) obtains the rough estimate of the depth. Next, they directly refined the depth using the isodepth constraint. Similar to Cui et al. [12], Yang et al. [13] used isodepth with simultaneous localization and mapping (SLAM) instead of SfM. SfM and SLAM both estimate the camera pose and the target geometry. However, SfM is often an offline process, whereas SLAM is often a real-time process. SfM usually consumes all images for calculation, and the camera pose can be discrete. SLAM usually uses a continuous movie, and assumes that the camera pose of the current frame, which is to be processed, has only a slight difference from the camera pose of the previous frame. Because SLAM can use the result of a previous frame to estimate the current frame, the method [13] can stably estimate the shape for multiple frames.

The photometric stereo method proposed by Mecca et al. [14] used the isodepth constraint obtained from polarization. Usually, the photometric stereo problem can be solved using three lights. If we have one constraint of isodepth, the photometric stereo problem can be solved with two lights. Calculating the ratio of two images under two lights can cancel the albedo parameter. Because there is no albedo parameter, the constraint of the surface normal obtained from two-light photometric stereo can be formulated as a linear equation. Additionally, the isodepth constraint can be formulated as a linear equation. These two types of constraints are concatenated for all pixels, resulting in a huge linear system. Note that the surface normal can be expressed by d_x and d_y, which is the differentiation of the depth, d. Therefore, the depth of the object is directly obtained as a closed-form solution of the linear system.

Fig. 7.11 Degree of polarization of diffuse reflection and thermal radiation

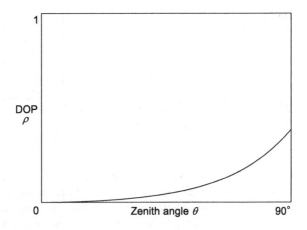

The surface normal has a strong relation to the degree of polarization. Candidates of zenith angle of the surface normal can calculated from the degree of polarization. Two candidates of zenith angle are obtained via specular reflection, whereas one candidate of the zenith angle is obtained from diffuse reflection. Two candidates of the azimuth angle of the surface normal are obtained from the phase angles 180° apart. Some methods solve the ambiguity of azimuth angle by propagating the azimuth angle from the occluding boundary to the inner part of the object area. There are some methods that solve the ambiguity of the zenith angle in specular reflection. Regarding the diffuse reflection, Miyazaki et al. [15] and Atkinson and Hancock [16] solved the ambiguity and estimated the shape. Regarding thermal radiation of infrared light, Miyazaki et al. [17] estimated the surface shape using polarization.

Figure 7.11 is a relationship between the zenith angle of the surface normal and the degree of polarization of diffuse reflection of an optically smooth surface, which is the same as the degree of polarization of thermal radiation.

$$\rho = \frac{1 - \cos^2 (\theta_1 - \theta_2)}{1 + \cos^2 (\theta_1 - \theta_2)}, \tag{7.3}$$

where Snell's law holds.

$$\sin \theta_1 = n \sin \theta_2 . \tag{7.4}$$

Here, the zenith angle θ is represented as the angle of emittance θ_1, and the index of refraction of the object is denoted as n. If the surface is rough, the degree of polarization will be lower than Eq. (7.3).

We have two candidates for estimating the zenith angle of a surface normal from the degree of polarization of specular reflection (Eq. (7.5) and Fig. 7.12).

$$\rho = \frac{\cos^2 (\theta_1 - \theta_2) - \cos^2 (\theta_1 + \theta_2)}{\cos^2 (\theta_1 - \theta_2) + \cos^2 (\theta_1 + \theta_2)} . \tag{7.5}$$

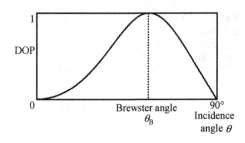

One candidate is true and the other is false. To solve the ambiguity of the zenith angle, we require additional data. Saito et al. [18] proposed a basic theory for estimating the surface normal of a transparent object using polarization. Barbour [19] approximated the relationship between the surface normal and the degree of polarization and developed a commercial sensor for shape from polarization. Miyazaki et al. [20] used a geometrical invariant to match the corresponding points from two views to estimate the surface normal of a transparent object. Huynh et al. [21] estimated the surface normal and index of refraction.

Reconstruction of shapes and appearances of thin film objects can be applied to many fields, including industrial inspection, biological analysis, and archaeologic research. Kobayashi et al. [22] proposed a method to estimate shapes and film thicknesses. Observed RGB values are represented by the integration of observed spectra (Eq. 7.6).

$$I_{RGB} = \int S_{RGB}(\lambda) R(\lambda) E(\lambda) d\lambda. \tag{7.6}$$

I_{RGB} is the observed RGB value. $S_{RGB}(\lambda)$ is the camera sensitivity function. $R(\lambda)$ and $E(\lambda)$ are reflectance and illumination spectra, respectively. The reflectance spectra of thin film, $R(\lambda)$, are defined as follows:

$$R(\lambda) = \left| \frac{r_{12} + r_{23} e^{i\Delta}}{1 + r_{23} r_{12} e^{i\Delta}} \right|^2. \tag{7.7}$$

r_{12} and r_{23} are reflectivities represented in Fresnel's law.

$$r_{12}^s = \frac{n_1 \cos \theta_1 - n_2 \cos \theta_2}{n_1 \cos \theta_1 + n_2 \cos \theta_2}, \tag{7.8}$$

$$r_{12}^p = \frac{n_2 \cos \theta_1 - n_1 \cos \theta_2}{n_2 \cos \theta_1 + n_1 \cos \theta_2}, \tag{7.9}$$

$$r_{23}^s = \frac{n_2 \cos \theta_2 - n_3 \cos \theta_3}{n_2 \cos \theta_2 + n_3 \cos \theta_3}, \tag{7.10}$$

$$r_{23}^p = \frac{n_3 \cos \theta_2 - n_2 \cos \theta_3}{n_3 \cos \theta_2 + n_2 \cos \theta_3}. \tag{7.11}$$

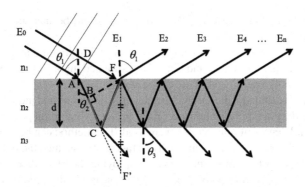

Fig. 7.13 Schematic of thin-film interference. n_1, n_2, and n_3 are indices of refraction of the incoming medium, thin film, and outgoing medium, respectively. θ_1 is zenith angle. θ_2 is transmitting angle. θ_3 is the angle of outgoing light transmitted in the thin film

Δ in Eq. (7.12) is phase difference.

$$\Delta = \frac{2\pi\phi}{\lambda},\qquad(7.12)$$

where ϕ is an optical path difference determined by the distance between points A and F in Fig. 7.13. The distance is $ACF - DF$. Considering a light entering a medium with an index of refraction, n_2, the speed of light in thin film is defined as $n_2 c$, where c is the light speed in the air. Therefore, the optical path difference becomes $n_2 ACF - n_1 DF$. In Fig. 7.13, n_1 is the index of refraction in the air. Thus, $n_1 = 1$. The optical path difference DF is equal to $n_2 AB$.

$$n_2 ACF - DF = n_2 BCF.\qquad(7.13)$$

The optical path difference, BCF, is equal to $BCF' = 2d \cos\theta_2$, because F' is a symmetrical point, F.

$$\phi = n_2 BCF = 2dn_2 \cos\theta_2.\qquad(7.14)$$

Therefore, in Eq. (7.6), the zenith angle, θ_1, indices of refraction, n_2 and n_3, and film thickness, d, are important parameters for appearances. In this case, only the film thickness is unknown and can be obtained from the lookup table.

Koshikawa and Shirai [23] used circular polarization to estimate the surface normal of a specular object. Mueller calculus was used to calculate the polarization state of the reflected light. However, extending their method to a dense estimation of a surface normal causes an ambiguity problem in which the surface normal cannot be uniquely determined. Guarnera et al. [24] extended their method to determine the surface normal uniquely, by changing the lighting conditions in two configurations. Morel et al. [25] also disambiguated it using multiple illuminations. However, they did not solve the ambiguity of the degree of polarization, because they did not use circular polarization.

As we described in Chap. 5, shading provides rich information about the surface normal. For example, Drbohlav and Sara [26] estimated an object shape using both the shape from polarization and photometric stereo.

Ngo et al. [27] estimated the surface normal, light direction, and index of refraction when both shading and polarization were used as clues. Regarding polarization, both phase angle and degree of polarization were used. The method can be applied even if the light-source directions are unknown. They randomly assumed the light directions and evaluated the cost function under them, employing the set of directions wherein the cost function was minimum. The optimization made the reconstruction error smaller, the surface normal smoother, and the variation of index of refraction smaller.

Usually, shape from polarization estimates the surface normal. Then, the depth can be integrated from the surface normal. However, Smith et al. [28] directly estimated the depth from polarization. The polarization of the diffuse reflection was analyzed, and both phase angle and degree of polarization were used. Shading information was also used as a constraint. A smoothness constraint was also added. To solve the convex/concave ambiguity, the convexity constraint was added. All of these constraints were represented as linear least squares formulations. As a result, the closed-form solution was obtained by using the QR solver implemented in the sparse matrix library.

The method proposed by Kadambi et al. [29] is known as *polarized 3D*. The depth obtained by Kinect gives a strong clue to the shape-from-polarization framework. Kinect gives a rough estimate of the surface normal, and the surface normal solves the ambiguity of the shape from polarization. The surface normal obtained by shape from polarization enhances the precision of the shape obtained by Kinect.

Shape-from-polarization methods often calculate the polarization parameters from input images first. Then, those methods estimate the surface normal from polarization parameters. They define the cost function to be minimized, either explicitly or implicitly, as the difference between input polarization image and output polarization image. However, Yu et al. [30] defined the cost function as the difference between the input images and the output images. Thus, the input images captured while rotating the linear polarizer were used as input images. Note that, if we know the height field of an object's surface, we can calculate its surface normal, and we can output the image under a specified angle of linear polarizer using the surface normal. Such input images and output images are used to form the cost function.

Miyazaki and Ikeuchi [31] solved the inverse problem of polarization ray tracing to estimate the surface normal of a transparent object. The algorithm of the polarization ray tracing method can be divided into two parts. First, the calculation of the propagation of the ray employed the same algorithm used in the conventional ray tracing method. Second, the calculation of the polarization state of the light applied a method called Mueller calculus [1–3]. We denote the input polarization image as $s_{\mathcal{I}}(x, y)$, where (x, y) represents the pixel position. Polarization ray tracing can render the polarization image from the shape of the transparent object. We denote this rendered polarization image as $s_{\mathcal{R}}(x, y)$. The shape of transparent objects is repre-

Fig. 7.14 Acquisition
system of transparent surface
measurement based on
polarization ray tracing

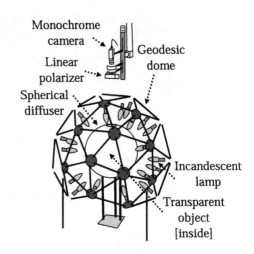

Monochrome
camera
Linear
polarizer
Spherical
diffuser
Geodesic
dome
Incandescent
lamp
Transparent
object
[inside]

sented as the height, $H(x, y)$, set for each pixel. Thus, the shape of the transparent
object is represented in 3D space, where the three axes are x, y, and H, and these
are Cartesian coordinates. Heights partially differentiated by the x- and y-axes are
called gradients, and are represented as p and q, respectively:

$$p = H_x = \frac{\partial H}{\partial x}, \quad q = H_y = \frac{\partial H}{\partial y}. \tag{7.15}$$

The surface normal, $\mathbf{n} = (-p, -q, 1)^\top$, is represented by these gradients. That is, a
surface normal is represented in 2D space, where the two axes are p and q, and this
space is called a gradient space. The rendered polarization image, $\mathbf{s}_\mathcal{R}(x, y)$, depends
upon height and surface normal. The problem is finding the best values to reconstruct
a surface, $H(x, y)$, that satisfy the following equation:

$$\mathbf{s}_\mathcal{I}(x, y) = \mathbf{s}_\mathcal{R}(x, y), \tag{7.16}$$

for all pixels (x, y). A definition of the cost function, which we want to minimize,
can be found:

$$\iint E_1(x, y)dxdy, \tag{7.17}$$

where

$$E_1(x, y) = \|\mathbf{s}_\mathcal{I}(x, y) - \mathbf{s}_\mathcal{R}(x, y)\|^2. \tag{7.18}$$

The target object is set inside the center of a plastic sphere with diameter 35-cm
(Fig. 7.14). This plastic sphere is illuminated by 36 incandescent lamps that are
almost uniformly distributed spatially around the plastic sphere using a geodesic
dome. The plastic sphere diffuses the light coming from the light sources, and it
behaves as a spherical light source that illuminates the target object from every

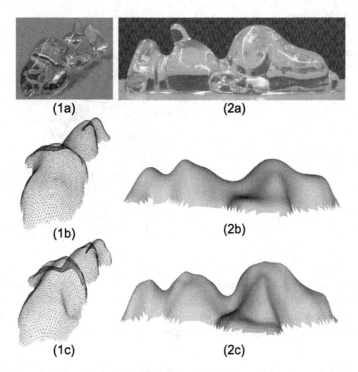

Fig. 7.15 Dog-shaped glass object. (1a)(2a) Real image. (1b)(2b) Initial state of the shape given manually. (1c)(2c) Estimated shape after 10 iterations

direction. The target object is observed using a monochrome camera from atop the plastic sphere through a 6-cm hole. A linear polarizer is set in front of the camera. The camera, object, and light sources are fixed. The least squares method calculates I_{max}, I_{min}, and ψ from four images taken by rotating the linear polarizer at 0°, 45°, 90°, and 135°. A dog-shaped object shown in Fig. 7.15(1a)(2a) is made of glass, and its index of refraction is 1.5. The shape obtained manually by human operation is used as an initial value for the experiment, as shown in Fig. 7.15(1b)(2b). The estimation results after 10 iterations are shown in Fig. 7.15 (1c)(2c).

7.2   Shape from Fluorescence

Fluorescence can be used for shape estimation. As is shown in Chap. 5, photometric stereo can be used to estimate the surface normal from the shading information. Shading phenomenon obeys Lambert's cosine law. Fluorescent light emits light from the surface, and its phenomenon is definitely different from diffuse reflection. The amount of light emitted from the surface is proportional to the amount of light absorbed at the surface. As can be seen in Fig. 7.16, the amount of light absorbed

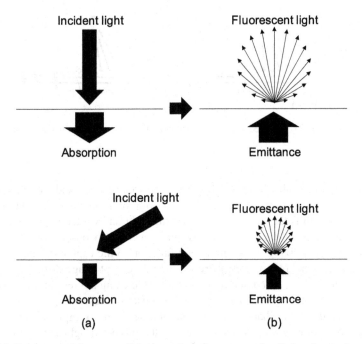

Fig. 7.16 Brightness of fluorescent light depends on the amount of radiation absorbed

obeys Lambert's cosine law. As a result, the brightness of the fluorescent light can be represented as Lambert's cosine law. Similar to photometric stereo, the surface normal can also be estimated from fluorescence.

The observed light can be a diffuse reflection if the observed wavelength is the same as the wavelength of the narrowband light. However, the observed light can be fluorescent emittance if the observed wavelength is different from the narrowband light, as is explained in Sect. 4.1.2. Therefore, we can separate the diffuse reflection and fluorescent light, as shown in Sect. 8.2.2. The fluorescent light is not a reflection but an emittance. Thus it does not cause specular reflection, because specular reflection is not an emittance but a reflection. As is shown in Chap. 5, photometric stereo *hates* specular reflection. Because the fluorescent light only causes diffuse reflection, it is suitable for surface normal estimation, as shown by Sato et al. [32] and Treibitz et al. [33].

Photometric stereo in murky water was proposed by Murez et al. [34] as a method to estimate the surface normal of the object that causes fluorescence, immersed in a scattering liquid. Suppose that we illuminate the object with a single color with a light of a single wavelength. The wavelength of reflected light on the object surface is the same as that of the incident light. Additionally, the wavelength of the scattering light is the same as those. To capture an image of reflected light, we require a sensor that detects the same wavelength as the reflected light. However, such a sensor also detects scattered light. However, the wavelength of fluorescent light and that of

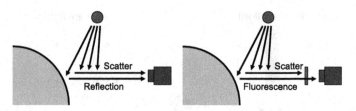

Fig. 7.17 Backscatter is removed, and the surface normal can be estimated from the fluorescence
of the object

incident light are usually different. Because the wavelength of incident light and that
of scattering light are same, if we set a filter in front of a camera to prevent the light
whose wavelength is the same as the scattering light and which transmits the light
whose wavelength is the same as fluorescent light, we only observe the fluorescent
light. After deconvolving the point-spread function to the captured image, we can
apply a conventional photometric stereo to the fluorescent image (Fig. 7.17).

Until now, we have discussed photometric stereo using fluorescent light. Hence-
forth, we introduce an active-stereo method that uses fluorescent light. Usually, it is
difficult to estimate the shape of a transparent object. For example, if we project a
line laser to an object with diffuse reflection, the projected light forms a curve on
the object's surface. The shape of a diffuse object can be estimated from the projected light using triangulation. However, a laser light will reflect and transmit at
the transparent object. Because the projected light will not appear on the surface of
a transparent object, conventional active stereo cannot be applied. Hullin et al. [35]
immersed the transparent object into a fluorescent liquid and fixed the object. They
projected a line laser that excited the fluorescence of the liquid, and the liquid became
bright at the area of light incidence. However, the transparent object remained dark,
because it did not cause fluorescence. Thus, the boundary between the bright and
dark areas became the object surface where the line laser was projected. Finally, the
shape was obtained easily from the boundary curve using the conventional active
stereo (Fig. 7.18).

7.3 Shape from Cast Shadow

The cast shadow of an object tells us the 3D geometry of that object's silhouette. There
are two key methods of reconstructing the shape from cast shadows. One recovers
the shape directly from the shadows of the target object [36]. Another recovers the
shape from the cast shadow on the target object by another object [37]. The former
has the advantage that no other devices are required. However, a shape-recovery
algorithm becomes complicated and unstable. The latter case requires an additional
object. However, the algorithm is usually simple. For example, the same algorithm of
light sectioning can be applied if a thin and straight rod is used as the occluder [37].

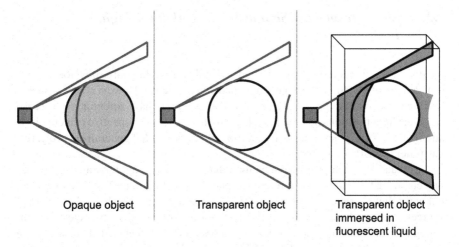

| Opaque object | Transparent object | Transparent object immersed in fluorescent liquid |

Fig. 7.18 Line-laser light is projected onto a target transparent object which is sunk in a fluorescent liquid

To recover the shape, the position of the light source should be known, and it is usually not an easy task. To mitigate the difficulty, techniques ignorant of the position have also been studied. Next, those studies are categorized into two types, based on with or without knowing the light position.

7.3.1 Shape from Cast Shadows with Known Light Positions

A shape from cast shadows with known light positions has a long history. Shafer and Kanade presented the mathematical formulation of shadow geometries and derived constraints for surface orientations from shadow boundaries [38]. Hambrick et al.proposed a method for classifying the boundaries of shadow regions [39]. Subsequently, several other methods for recovering 3D shapes up to Euclidean reconstructions based on geometrical constraints of cast shadows were proposed [40–43]. Yu and Chang [44] reconstructed the height field of bumpy textures under an assumption of known light directions. All these methods assumed that objects casting shadows were static and that the light directions or positions were known. Yamazaki et al. [45] proposed a method for 3D reconstruction by applying a shape-from-silhouette technique to cast shadows projected onto a single plane and were able to obtain accurate results. Although the light positions were not calibrated in advance, they used shadows of spheres captured with the shadows of the target object to estimate the light positions for each of the multiple images.

7.3.2   Shape from Cast Shadows with Unknown Light Positions

If a Euclidean shape can be reconstructed with unknown light-source positions, it may broaden the application of shape-from-shadow techniques. However, it was proven that scene reconstructions based on binary shadow regions have ambiguities for 4 DoF when the light positions are unknown [46]. In the case of a perspective camera, these ambiguities correspond to a family of transformations called generalized projective bas-relief (GPBR) transformations.

Caspi and Werman proposed a method that used two straight, parallel, and fixed objects to cast shadows and a reference plane (e.g., the ground) [47] to deal with unknown light-source positions. To solve ambiguities caused by unknown light sources, they used parallelisms of the shadows of straight edges by detecting vanishing points. Kawasaki and Furukawa proposed a more general method, where the camera was completely uncalibrated [37]. Thus, the occluder and the light source can be moved, and the light source can be a parallel or point light source. Wider constraint types than parallelisms of shadows can be used to resolve ambiguities. They used several types of coplanar constraints for reconstruction. Details of the technique are briefly explained in next.

Shape from Coplanarity of Cast Shadows

If a set of points exist on the same plane, they are said to be coplanar. Since a scene comprising plane structures has many coplanarities, a coplanarity that is actually observed as a real plane in the scene is termed an *explicit coplanarity* in their method. Thus, points or curves on the real plane are said to be *explicit-coplanar*; and a real plane is sometimes referred to as an *explicit plane*. All points in the region of an image corresponding to a planar surface are explicit-coplanar. For example, in Fig. 7.19a, the sets of points on a region of the same color, except for white, are explicit-coplanar.

Conversely, in a 3D space, there exist an infinite number of coplanarities that are not explicitly observed in ordinary situations. However, they can be observed under specific conditions. For example, the boundary of a cast shadow of a straight edge is the set of coplanar points shown in Fig. 7.19b. This kind of coplanarity is not visible until the shadow is cast on the scene; these coplanarities are termed *implicit coplanarities*. Points or curves on the shadow boundary are said to be *implicit-coplanar*, and a plane that includes implicit-coplanar points is referred to as an *implicit plane*. Implicit coplanarities can be observed in various situations, such as in the case where buildings having straight edges under the sun cast shadows onto a scene. Although explicit-coplanar curves are observed only for planar parts of a scene, implicit coplanarities can be observed on arbitrary-shaped surfaces.

For shape reconstruction, first, points on the intersections of multiple planes are extracted from the image of the scene. These planes may be explicit or implicit. From the locations of the points on the images, simultaneous linear equations can be created. By solving the simultaneous equations, the scene can be reconstructed up to a 4-DoF indeterminacy, which is an essential property of the simultaneous equations, and another indeterminacy is caused by unknown camera parameters. The indeterminacies of the solution are represented as 3D homographies, called *projective reconstruction*. To obtain a Euclidean reconstruction from the solution, those indeterminacies must be upgraded by using constraints other than coplanarities as follows:

(1) In Fig. 7.19c, the ground is plane π_0, and the linear object λ stands vertically on the ground. If the planes corresponding shadows of λ are π_1 and π_2, $\pi_0 \perp \pi_1$ and $\pi_0 \perp \pi_2$ can be derived from $\lambda \perp \pi_0$.

(2) In the same figure, the sides of box B are π_3, π_4, and π_5. If box B is rectangular, π_3, π_4, and π_5 are orthogonal to each other. If box B is on the ground, π_3 is parallel to π_0.

From the constraints available from the scene, such as the above examples, variables corresponding to the remaining indeterminacy can be determined and metric reconstruction can be achieved. With enough constraints, the camera parameters can be estimated at the same time. The actual flow of the algorithm is as follows.

- **Extraction of coplanarity constraints**. From a series of images that are acquired from a scene with shadows captured by a fixed camera, shadow boundaries are extracted as implicit-coplanar curves. If the scene has plane areas, explicit-coplanar points are sampled from the area. Here, explicit-coplanar curves and points are acquired by simply drawing curves on a planar region of the image as shown in Fig. 7.19d.
- **Projective reconstruction using coplanarity constraints** . Constraints are acquired as linear equations from a dataset of coplanarities. By numerically solving the simultaneous equations, projective reconstruction with indeterminacy of 4+ DoFs can be acquired.
- **Metric reconstruction using metric constraints** . To achieve metric reconstruction, an upgrade process of the **projective reconstruction** is required. The solution can be upgraded by solving the metric constraints.
- **Dense-shape reconstruction**. The processes of **projective reconstruction** and **metric reconstruction** are performed on selected frames. To realize dense-shape reconstruction of a scene, implicit-coplanar curves from all images are used to reconstruct 3D shapes using the triangulation method shown in Fig. 7.20.

Shape reconstruction results obtained from an indoor scene using a point light source are shown in Fig. 7.21. A video camera was directed toward the target object and multiple boxes, and the scene was captured while the light source and the bar for shadowing moved freely. From the captured image sequence, several images were selected and the shadow curves of the bar were detected. By using the detected coplanar shadow curves, 3D reconstruction up to 4-DoF indeterminacy was performed. Metric reconstruction was performed up to scale using the nonlinear method. For the metric reconstruction, orthogonality of the faces of the boxes was used. Since there were only small noises because of the indoor environment, a dense and accurate shape was correctly reconstructed.

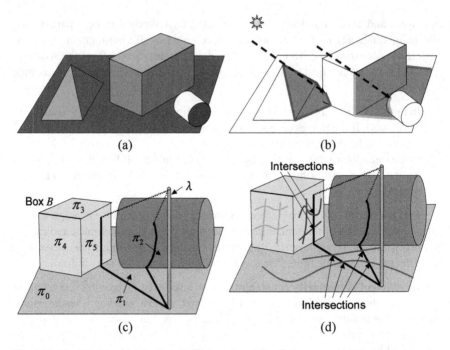

(a)   (b)

(c)   (d)

Fig. 7.19 Coplanarities in a scene: **a** Explicit coplanarities. Points on a region of the same color except for white are a set of explicit-coplanar points. Note that the points on a region of a curved surface are not coplanar. **b** Implicit coplanarities. Lines of each color are implicit-coplanar. **c** Examples of metric constraints: π_1 and π_2 are implicit planes and π_0, π_3, π_4, and π_5 are explicit planes. $\pi_0 \perp \pi_1$ and $\pi_0 \perp \pi_2$ if $\lambda \perp \pi_0$. $\pi_3 \perp \pi_4$, $\pi_4 \perp \pi_5$, $\pi_3 \perp \pi_5$, and π_3 and π_0 are parallel if box B is rectangular and is placed on π_0. **d** Intersections between explicit-coplanar curves and implicit-coplanar curves in a scene. Curves of each color are within a planar region and they are explicit-coplanar. Explicit-coplanar curves can be drawn by hand

Fig. 7.20 Triangulation method

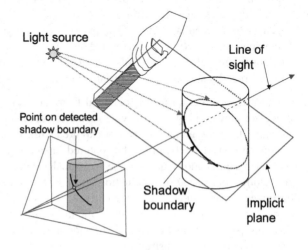

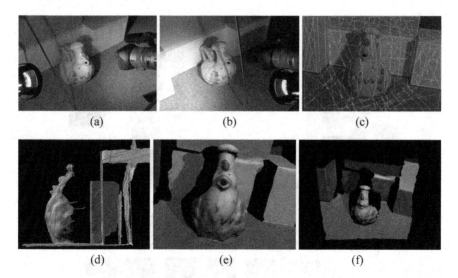

(a)   (b)   (c)

(d)   (e)   (f)

Fig. 7.21 Reconstruction of a real indoor scene: **a, b** Captured frames. **c** Implicit- and explicit-coplanar curves and their intersection points. **d** Reconstructed coplanar shadow curves (red) with a dense reconstructed model (shaded). **e, f** Textured reconstructed model

7.4 Shape from Projector Defocus

Depth from defocus (DfD) is a popular technique used to recover a shape from a single image captured by a camera. Likewise, the technique can also be applied to projector defocus. Moreno-Noguer et al. used a small circular aperture to realize DfD [48] and Kawasaki et al. add a coded aperture (CA) on a video projector to improve the accuracy and density of DfD [49, 50]. As shown in Fig. 7.22, Moreno-Noguer et al. projected dense grid points and observed the sizes of the dots to estimate depths

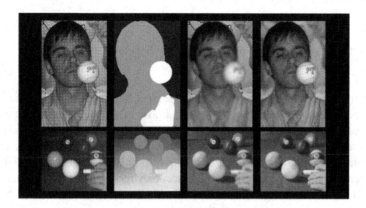

Fig. 7.22 Depth from defocus (DfD) using standard aperture shape (circular)

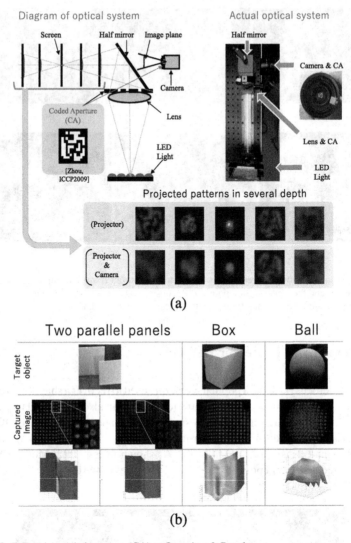

Fig. 7.23 DfD using coded aperture (CA). **a** Overview. **b** Results

from their diameters. Because it is difficult to achieve high accuracy by just measuring diameters of small and blurry circles, the purpose of the technique was not depth measurement, but a rough segmentation of the scene. For a solution, Kawasaki et al.proposed a technique using a CA and estimated the depth of the scene using a deconvolution-based approach [49, 50], as shown in Fig. 7.23a. By optimizing the CA pattern using genetic algorithm, high-accuracy reconstruction was achieved, as shown in Fig. 7.23. However, for all techniques, depth range was limited, because defocus blur increased rapidly with the depth change.

Fig. 7.24 Optical configuration. d_f is a depth of the focal plane

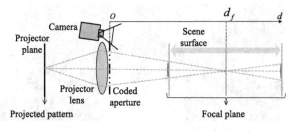

Fig. 7.25 Actual optical system. **a** Setup with a short baseline. **b** CA installed on the projector lens

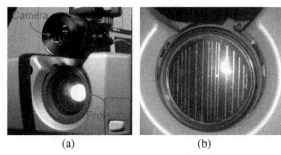

(a) (b)

Fig. 7.26 Depth estimation results. Brown broken line represents DfD result [49]; green dotted line represents simple random dot stereo result; yellow dotted line represents the result using a circular aperture with the same algorithm; and red dotted line represents the result using a ToF sensor (Kinect v2). Note that the size of the circular aperture is equal to the total size of the slit aperture

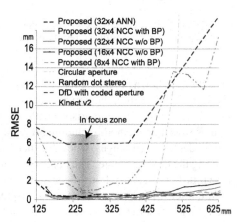

To solve the narrow depth range problem, a novel light-field projector comprising an off-the-shelf video projector and a CA mask attached was proposed [51]. The system has the same capability as a densely arranged array of projectors, and the projected patterns change their appearance depending on depth. This contrasts with the traditional DfD projection system, where generated patterns are just scale differences with respect to depth. By using depth-dependent patterns, camera and projector are not required to be set up with a wide baseline. Even with no baseline, the depth can still be reconstructed. The system setup is similar to the common active-stereo setup shown in Figs. 7.24 and 7.25a. A projector and a camera are placed at a certain baseline, and a light pattern is projected on the object. The difference from conventional systems is the CA placed over the projector lens to realize a light field

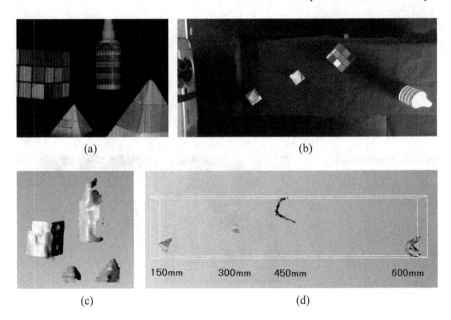

Fig. 7.27 Wide depth range shape reconstruction result. **a** Input image. **b** Top view. **c** Reconstructed result (side view). **d** Reconstructed result (top view)

projection as shown in Fig. 7.25b. The results of using the optical system are shown in Fig. 7.26. In the Figure, the depth value was estimated by using Kawasaki's method and other methods for comparisons with root-mean-square error (RMSE). As can be seen, Kawasaki's method, including all window sizes, the ToF sensor recovered the correct depths of all ranges, whereas others rapidly decreased their accuracies when entering the defocus range. The depth estimation results of more generic objects are shown in Fig. 7.27a, b. Four objects are placed about 150, 300, 450, and 620 mm from the lens. Figure 7.27a shows the captured image with the projected pattern, and (c) and (d) show the reconstruction results. It is confirmed that the shapes were correctly estimated at the right position with fine details using Kawasaki's method.

References

1. Born M, Wolf E (2013) Principles of optics: electromagnetic theory of propagation, interference and diffraction of light. Elsevier
2. Hecht E (1998) Hecht optics vol 997. Addison Wesley, pp 213–214
3. Shurcliff WA (1966) Polarized light. Production and use. Harvard University Press, Cambridge, MA1 c1966
4. Wolff LB, Boult TE (1991) Constraining object features using a polarization reflectance model. IEEE Trans Pattern Anal Mach Intell (PAMI) 7:635–657

5. Rahmann S (1999) Inferring 3d scene structure from a single polarization image. In: Polarization and color techniques in industrial inspection, vol 3826. International Society for Optics and Photonics, pp 22–34

6. Rahmann S (2000) Polarization images: a geometric interpretation for shape analysis. In: Proceedings of the international conference on pattern recognition (ICPR), vol 3. IEEE, pp 538–542

7. Rahmann S, Canterakis N (2001) Reconstruction of specular surfaces using polarization imaging. In: Proceedings of the IEEE conference on computer vision and pattern recognition (CVPR), vol 1. IEEE, pp I–I

8. Rahmann S (2003) Reconstruction of quadrics from two polarization views. In: Iberian conference on pattern recognition and image analysis. Springer, pp 810–820

9. Atkinson GA, Hancock ER (2007) Shape estimation using polarization and shading from two views. IEEE Trans Pattern Anal Mach Intell (PAMI) 29(11):2001–2017

10. Miyazaki D, Shigetomi T, Baba M, Furukawa R, Hiura S, Asada N (2016) Surface normal estimation of black specular objects from multiview polarization images. Opt Eng 56(4):041303

11. Miyazaki D, Furuhashi R, Hiura S (2020) Shape estimation of concave specular object from multiview polarization. J Electron Imaging 29(4):041006

12. Cui Z, Gu J, Shi B, Tan P, Kautz J (2017) Polarimetric multi-view stereo. In: Proceedings of the IEEE conference on computer vision and pattern recognition (CVPR), pp 1558–1567

13. Yang L, Tan F, Li A, Cui Z, Furukawa Y, Tan P (2018) Polarimetric dense monocular slam. In: Proceedings of the IEEE conference on computer vision and pattern recognition (CVPR), pp 3857–3866

14. Mecca R, Logothetis F, Cipolla R (2018) A differential approach to shape from polarization

15. Miyazaki D, Tan RT, Hara K, Ikeuchi K (2003) Polarization-based inverse rendering from a single view. In: Proceedings of the international conference on computer vision (ICCV). IEEE Computer Society, p 982

16. Atkinson GA, Hancock ER (2006) Recovery of surface orientation from diffuse polarization. IEEE Trans Image Process (TIP) 15(6):1653–1664

17. Miyazaki D, Saito M, Sato Y, Ikeuchi K (2002) Determining surface orientations of transparent objects based on polarization degrees in visible and infrared wavelengths. J Opt Soc Am A 19(4):687–694

18. Saito M, Sato Y, Ikeuchi K, Kashiwagi H (1999) Measurement of surface orientations of transparent objects by use of polarization in highlight. J Opt Soc Am A 16(9):2286–2293

19. Barbour BA (2012) Apparatus and method for extracting information from electromagnetic energy including target 3d structure and materials. US Patent 8,320,661

20. Miyazaki D, Kagesawa M, Ikeuchi K (2004) Transparent surface modeling from a pair of polarization images. IEEE Trans Pattern Anal Mach Intell (PAMI) 1:73–82

21. Huynh CP, Robles-Kelly A, Hancock ER (2013) Shape and refractive index from single-view spectro-polarimetric images. Int J Compu Vis (IJCV) 101(1):64–94

22. Kobayashi Y, Morimoto T, Sato I, Mukaigawa Y, Tomono T, Ikeuchi K (2016) Reconstructing shapes and appearances of thin film objects using RGB images. In: Proceedings of the IEEE conference on computer vision and pattern recognition (CVPR), pp 3774–3782

23. Koshikawa K, Shirai Y (1987) A model-based recognition of glossy objects using their polarimetrical properties. Adv Robot 2(2):137–147

24. Guarnera GC, Peers P, Debevec P, Ghosh A (2012) Estimating surface normals from spherical stokes reflectance fields. In: Proceedings of the European conference on computer vision (ECCV). Springer, pp 340–349

25. Morel O, Stolz C, Meriaudeau F, Gorria P (2006) Active lighting applied to three-dimensional reconstruction of specular metallic surfaces by polarization imaging. Appl Opt 45(17):4062–4068

26. Drbohlav O, Sara R (2001) Unambiguous determination of shape from photometric stereo with unknown light sources. In: Proceedings of the international conference on computer vision (ICCV), vol 1. IEEE, pp 581–586

27. Ngo Thanh T, Nagahara H, Taniguchi RI (2015) Shape and light directions from shading and polarization. In: Proceedings of the IEEE conference on computer vision and pattern recognition (CVPR), pp 2310–2318

28. Smith W, Ramamoorthi R, Tozza S (2018) Height-from-polarisation with unknown lighting or albedo. In: IEEE transactions on pattern analysis and machine intelligence (PAMI)

29. Kadambi A, Taamazyan V, Shi B, Raskar R (2017) Depth sensing using geometrically constrained polarization normals. Int J Comput Vis (IJCV) 125(1–3):34–51

30. Yu Y, Zhu D, Smith WA (2017) Shape-from-polarisation: a nonlinear least squares approach. In: Proceedings of the international conference on computer vision (ICCV), pp 2969–2976

31. Miyazaki D, Ikeuchi K (2007) Shape estimation of transparent objects by using inverse polarization ray tracing. IEEE Trans Pattern Anal Mach Intell (PAMI) 29(11):2018–2030

32. Sato I, Okabe T, Sato Y (2012) Bispectral photometric stereo based on fluorescence. In: Proceedings of the IEEE conference on computer vision and pattern recognition (CVPR). IEEE, pp 270–277

33. Treibitz T, Murez Z, Mitchell BG, Kriegman D (2012) Shape from fluorescence. In: Proceedings of the European conference on computer vision (ECCV). Springer, pp 292–306

34. Murez Z, Treibitz T, Ramamoorthi R, Kriegman D (2015) Photometric stereo in a scattering medium. In: Proceedings of the international conference on computer vision (ICCV), pp 3415–3423

35. Hullin MB, Fuchs M, Ihrke I, Seidel HP, Lensch HP (2008) Fluorescent immersion range scanning. ACM Trans Graph (TOG) 27(3):87

36. Okabe T, Sato I, Sato Y (2009) Attached shadow coding: estimating surface normals from shadows under unknown reflectance and lighting conditions. In: Proceedings of the international conference on computer vision (ICCV). IEEE, pp 1693–1700

37. Kawasaki H, Furukawa R (2007) Shape reconstruction from cast shadows using coplanarities and metric constraints. In: Proceedings of the Asian conference on computer vision (ACCV) LNCS 4843, vol II, pp 847–857

38. Shafer SA, Kanade T (1983) Using shadows in finding surface orientations. Comput Vis Graph Image Process 22(1):145–176

39. Hambrick LN, Loew MH, Carroll RLJ (1987) The entry exit method of shadow boundary segmentation. IEEE Trans Pattern Anal Mach Intell (PAMI) 9(5):597–607

40. Hatzitheodorou M, Kender J (1988) An optimal algorithm for the derivation of shape from shadows. In: Proceedings of the IEEE conference on computer vision and pattern recognition (CVPR), pp 486–491

41. Raviv D, Pao Y, Loparo KA (1989) Reconstruction of three-dimensional surfaces from two-dimensional binary images. IEEE Trans Robot Autom 5(5):701–710

42. Daum M, Dudek G (1998) On 3-d surface reconstruction using shape from shadows. In: Proceedings of the IEEE conference on computer vision and pattern recognition (CVPR), pp 461–468. citeseer.ist.psu.edu/daum98surface.html

43. Savarese S, Andreetto M, Rushmeier H, Bernardini F, Perona P (2007) 3D reconstruction by shadow carving: theory and practical evaluation. Int J Comput Vis (IJCV) 71(3):305–336. https://doi.org/10.1007/s11263-006-8323-9

44. Yu Y, Chang JT (2005) Shadow graphs and 3D texture reconstruction. Int J Comput Vis (IJCV) 62(1–2):35–60. https://doi.org/10.1007/s11263-005-4634-5

45. Yamazaki S, Narasimhan SG, Baker S, Kanade T (2007) Coplanar shadowgrams for acquiring visual hulls of intricate objects. In: Proceedings of the international conference on computer vision (ICCV)

46. Kriegman DJ, Belhumeur PN (2001) What shadows reveal about object structure. J Opt Soc Am 18(8):1804–1813

47. Caspi Y, Werman M (2003) Vertical parallax from moving shadows. In: Proceedings of the IEEE conference on computer vision and pattern recognition (CVPR), vol 2. IEEE, pp 2309–2315

48. Nayar S, Watanabe M, Noguchi M (1996) Real-Time focus range sensor. IEEE Trans Pattern Anal Mach Intell (PAMI) 18(12):1186–1198

49. Kawasaki H, Horita Y, Masuyama H, Ono S, Kimura M, Takane Y (2013) Optimized aperture for estimating depth from projector's defocus. In: 2013 International conference on 3d vision-3DV 2013. IEEE, pp 135–142
50. Kawasaki H, Horita Y, Morinaga H, Matugano Y, Ono S, Kimura M, Takane Y (2012) Structured light with coded aperture for wide range 3d measurement. In: 2012 19th IEEE international conference on image processing. IEEE, pp 2777–2780
51. Kawasaki H, Ono S, Horita Y, Shiba Y, Furukawa R, Hiura S (2015) Active one-shot scan for wide depth range using a light field projector based on coded aperture. In: Proceedings of the international conference on computer vision (ICCV), pp 3568–3576

Chapter 8
Photometric Estimation

Previous chapters considered methods of estimating the geometric shape information by using active-lighting methods, which provide photometric characteristics of
the target scene and light sources. Because BRDF depends on materials, adequate
lighting and analysis provides rich information about those materials. Moreover,
multispectral light gives us a more robust identification capability. Furthermore,
polarization and fluorescence are useful for material classification. Analyzing captured images enables us to separate an image into several photometric components.
Additionally, a scene illuminated by multiple lights can be decomposed into an
image illuminated by a single light. As such, active lighting is useful, not only for
geometrical analysis, but also for photometrical analysis.

8.1 Material Estimation from Spectra

Multispectral images have richer information than an RGB image and supports a wide
variety of applications. Different materials often have different spectral reflectances.
In this section, we explain some methods that classify each material using multispectral analysis.

First, there is a method of image segmentation that uses a multispectral image.
Let $\mathcal{I} = \{\mathbf{I}_1, \mathbf{I}_2, \mathbf{I}_3, \ldots, \mathbf{I}_i, \ldots, \mathbf{I}_N\}$, where $\mathbf{I}_i$ is input data, of m dimensions at node
i of a graph structure. Then, the normal-cuts (NCuts) graph-cut method calculates
the weight matrix, $\mathbf{W}$, representing similarity among nodes:

$$W_{ij} = \tilde{W}_{ij} \hat{W}_{ij} , \tag{8.1}$$

$$\tilde{W}_{ij} = \exp\left(\frac{-\|\mathbf{I}_i - \mathbf{I}_j\|^2}{\sigma_I^2}\right) ,$$

$$\hat{W}_{ij} = \begin{cases} \exp\left(\frac{-\|\mathbf{X}_i - \mathbf{X}_j\|_2^2}{\sigma_X^2}\right) & (\text{if } \|\mathbf{X}_i - \mathbf{X}_j\|_2 < r) \\ 0 & (\text{otherwise}) \end{cases} ,$$

© Springer Nature Switzerland AG 2020
K. Ikeuchi et al., *Active Lighting and Its Application for Computer Vision*,
Advances in Computer Vision and Pattern Recognition,
https://doi.org/10.1007/978-3-030-56577-0_8

where $\mathbf{I}_i$ and $\mathbf{I}_j$ are input values of m dimensions at nodes i and j. σ_I^2 is the variance of input data, and r is the threshold of the proximity between the two nodes in the image. Additionally, $\mathbf{X}_i$ and $\mathbf{X}_j$ are pixel positions, and σ_X^2 is its variance. NCuts solves the generalized eigensystem equation,

$$(\mathbf{D} - \mathbf{W})\mathbf{y} = \lambda \mathbf{D}\mathbf{y}, \tag{8.2}$$

where $\mathbf{D}$ is an $N \times N$ diagonal matrix, $\mathbf{D} = \mathrm{diag}(W_{1,1}, W_{2,2}, \ldots, W_{N,N})$, and $\mathbf{W}$ is an $N \times N$ symmetric matrix, $W_{ij} = W_{ji}$. Next, the Laplacian matrix, $\mathbf{L} = (\mathbf{D} - \mathbf{W})$, can be calculated from the weight matrix, $\mathbf{W}$. The normalized Laplacian matrix, $\tilde{\mathbf{L}}$, is given by

$$\tilde{\mathbf{L}} = \mathbf{I} - \mathbf{D}^{-\frac{1}{2}}\mathbf{W}\mathbf{D}^{-\frac{1}{2}}, \tag{8.3}$$

We can transform Eq. (8.2) into a standard eigensystem, as follows:

$$\mathbf{D}^{-\frac{1}{2}}\mathbf{W}\mathbf{D}^{-\frac{1}{2}}\mathbf{z} = (1 - \lambda)\mathbf{z}. \tag{8.4}$$

We can span a low-dimensional space of E dimensions with the eigenvectors from the $E + 1$ least-significant eigenvalues, where E is the partition number. We ignore the least-significant eigenvalue and the corresponding eigenvector. In the least significant space, all input data have roughly the same values because of data normalization. We map the input data onto this low-dimensional space. Finally, we segment them into E clusters using the k-means method. This method was first proposed by Morimoto and Ikeuchi [1]. Figure 8.1 shows a panoramic multispectral image. Figure 8.2 shows its segmentation result.

As shown, multispectral analysis can identify target materials. A multispectral light is beneficial for this purpose. Blasinski et al. [2] used a support vector classifier to design the illumination needed to distinguish two materials. They illuminated an object using three types of multispectral light one-by-one and captured the object with a monochrome camera. The material of each pixel was distinguished from those images.

The measurement apparatus of Skaff et al. [3] formed a dome surrounding the target object. Six types of LEDs were placed at 25 positions on the dome. The researchers identified the material from the monochrome image of the object, which was illuminated under multispectral lighting conditions. When classifying each material, it is sometimes redundant to use 150 images of 6 types and 25 orientations. Yet, their algorithm preliminarily studied a small amount of data to classify two materials.

The device developed by Sato et al. [4] identified each material in real time. They compiled five LED types as one cluster, and they put four clusters at different points on the sensor. Features of surface quality, average pixel brightness, and darkest and brightest pixel brightness values were calculated from the input data obtained by the device. They used these four parameters to train a support vector machine (SVM) and classified each material.

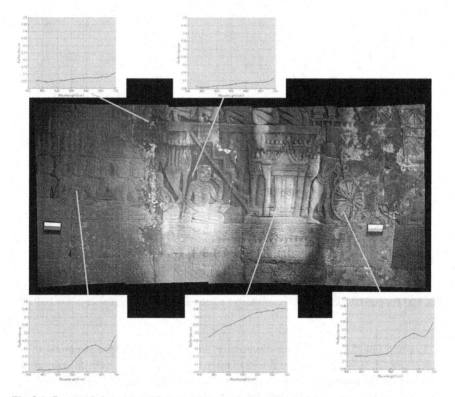

Fig. 8.1 Panoramic image made from multispectral images. This image has 81-dimensional spectra at each pixel

Fig. 8.2 Segmentation result of panoramic multispectral image

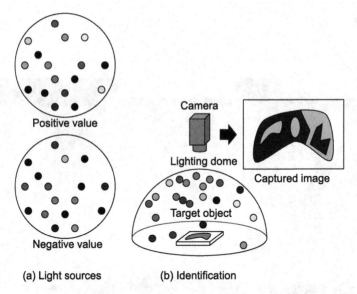

Fig. 8.3 Per-pixel classification of raw materials based on optimal projections of spectral BRDF

Liu and Gu [5] designed an illuminator that classified surface materials of an object (Fig. 8.3). Because spectral BRDF often differs when the material is different, we can classify materials using multispectral lights. They placed them on the dome structure, which was placed over the target object. Suppose that we want to distinguish a certain material, P, and another material, N. Depending on the spectral BRDF of the materials, P strongly reflects light in a certain direction at a certain wavelength, and material N reflects less in that direction and wavelength. Furthermore, if we choose another direction and wavelength, material N strongly reflects while material P weakly reflects. Therefore, if the multispectral lights are set on the dome adequately, the appearances of materials P and N will differ enough for analysis. Using training data, the brightness of each light is learned using an SVM so that materials P and N can be distinguished. The discriminative vector obtained by SVM distinguishes the two materials, and each element of the vector represents the brightness of light, which is set according to the vector. The target material is illuminated under the dome. Then, the captured image provides the classification result. The discriminative vector has partial negative values. Therefore, the light of a positive value is lit first, and the light of negative value is lit next, where the absolute value is used as the brightness (Fig. 8.3a). Subtracting the image under negative light from the image under positive light can represent the negative values of the discriminative vector. This subtracted image becomes the segmentation result (Fig. 8.3b). For example, if the pixel value is high, it is material P, and, if the pixel value is low, it is material Wang and Okabe [6] forced the weight (brightness) of each measure (light) to be positive during the training stage. Unlike Liu and Gu [5], who required a pair of images, Wang and Okabe [6] identified the material from a single image.

8.2 Material Estimation from Polarization and Fluorescence

8.2.1 Material Estimation from Polarization

Polarization is very useful for material estimation, because, when it is reflected on an object's surface, it changes its visible state. Dielectric materials strongly polarize light, and metals weakly polarize it. Therefore, we can distinguish dielectrics and metals via polarization. According to Fresnel's law, the intensity reflectivity is represented as Eqs. (8.5) and (8.6).

$$R_p = R_s \frac{a^2 + b^2 - 2a \sin\theta \tan\theta + \sin^2\theta \tan^2\theta}{a^2 + b^2 + 2a \sin\theta \tan\theta + \sin^2\theta \tan^2\theta}, \tag{8.5}$$

$$R_s = \frac{a^2 + b^2 - 2a \cos\theta + \cos^2\theta}{a^2 + b^2 + 2a \cos\theta + \cos^2\theta}, \tag{8.6}$$

$$2a^2 = \sqrt{\left(n^2 - k^2 - \sin^2\theta\right)^2 + 4n^2k^2} + \left(n^2 - k^2 - \sin^2\theta\right),$$

$$2b^2 = \sqrt{\left(n^2 - k^2 - \sin^2\theta\right)^2 + 4n^2k^2} - \left(n^2 - k^2 - \sin^2\theta\right),$$

$$a \geq 0, \quad b \geq 0,$$

$$\eta = n - ik.$$

Here, R_p is the reflectivity parallel to the reflection plane, and R_s is the perpendicular. The index of refraction, η, of metal is a complex number, where i is the imaginary component. Parameter n represents the index of refraction, and k is the coefficient of extinction. Figure 8.4 is the graph of R_p and R_s for the dielectric with $\eta = 1.7$. Figure 8.5 displays metal with $\eta = 0.82 - 5.99i$. The degree of polarization $(R_s - R_p)/(R_s + R_p)$ is shown in Fig. 8.6. As can be seen, the degree of polarization is high for dielectrics and low for metals. The difference can be enhanced if we use this ratio:

$$\frac{R_s}{R_p}. \tag{8.7}$$

From Fig. 8.7, the ratio is nearly 1 for metals but is larger for dielectrics. Wolff and Boult [7] distinguished metals and dielectrics in polarization imagery for each pixel.

Martinez-Domingo et al. [8] used both polarization and multispectral analysis to segment regions of materials. A mean-shift algorithm and a multispectral image were used for region segmentation. They classified each region as metal or dielectric using the curvature of the polarization degree. Otherwise, the ratio of reflectivity of linear polarization can be used to identify materials. Chen and Wolff [9], however, used the phase retardation. The phase shift is represented as follows [10].

Fig. 8.4 Intensity reflectance of dielectric medium ($\eta = 1.7$)

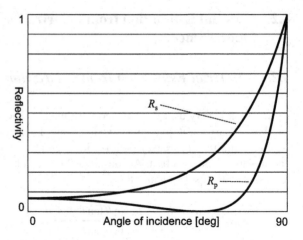

Fig. 8.5 Intensity reflectance of metallic medium ($\eta = 0.82 - 5.99i$)

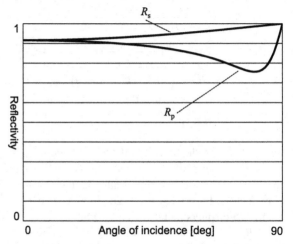

$$\tan \delta_\parallel = \frac{2 \cos \psi [(n^2 - k^2)t - (2nk)s]}{(n^2 + k^2)^2 \cos^2 \psi - (s^2 + t^2)} , \tag{8.8}$$

$$\tan \delta_\perp = \frac{2t \cos \psi}{\cos^2 \psi - (s^2 + t^2)} . \tag{8.9}$$

$$2s^2 = \sqrt{(n^2 - k^2 - \sin^2 \psi)^2 + 4n^2k^2} + (n^2 - k^2 - \sin^2 \psi) ,$$

$$2t^2 = \sqrt{(n^2 - k^2 - \sin^2 \psi)^2 + 4n^2k^2} - (n^2 - k^2 - \sin^2 \psi) .$$

Here, $n - ik$ is the index of refraction, and i is the imaginary number. For the dielectric material, $k = 0$. For metal, $k \neq 0$. Relative phase shift can be expressed as

Fig. 8.6 Degree of polarization of dielectric and metal

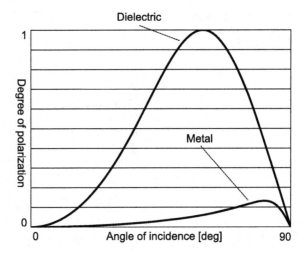

Fig. 8.7 Ratio of reflectance of dielectric and metal

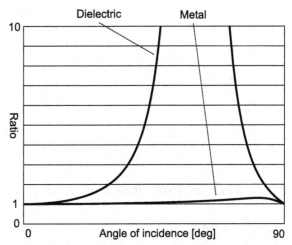

$\delta = \delta_{\parallel} - \delta_{\perp}$. Equations (8.8) and (8.9) show that $\delta = 0$ for dielectric material and $\delta \neq 0$ for metal (Fig. 8.8). A linear polarizer solely cannot measure δ, but adding a quarter-wave plate retarder results in success. Chen and Wolff [9] mathematically proved that, with a retarder, the phase angle, θ_{max} (Fig. 8.9), becomes 0° or 90° for dielectrics and other values for metals. Therefore, measuring θ_{max} identifies the material.

Fig. 8.8 Phase retardance (index of refraction $0.82 - 5.99i$)

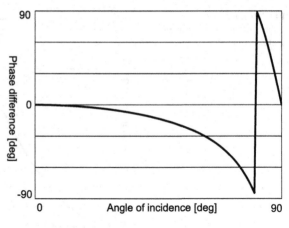

Fig. 8.9 Phase shift

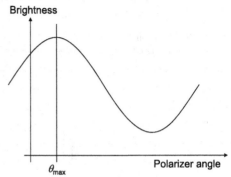

8.2.2   Material Estimation from Fluorescence

The color of the fluorescent light depends on its constituent materials. Thus, if we know its color, we can identify the fluorescent material. Usually, both reflection and fluorescence are observed at the same time. Separating these gives us the region material segmentation with and without fluorescent material. The separated fluorescent component can be used for photometric stereo (Chap. 5 and Sect. 7.2).

Reflection and fluorescence are different phenomena, but their compositions are observed simultaneously. To separate them, the input data should be captured with a multispectral camera or an RGB color camera. These can both obtain multichannel images at different wavelengths. Incident light is reflected at a certain scale and wavelength, whereas other wavelengths are often emitted fluorescently. Although the input image is a composition of two parts, the wavelength dependency is different between reflection and fluorescence. Thus, independent component analysis can decompose them [11] (Fig. 8.10).

Decomposition will be robust if the input image is multispectral. The method proposed by Fu et al. [12] represented fluorescence as chromaticity and reflection as

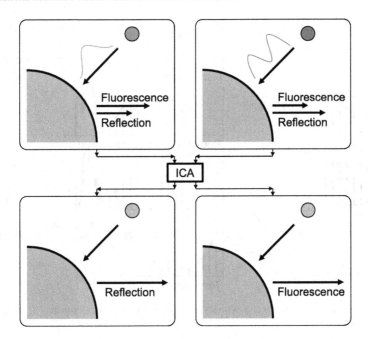

Fig. 8.10 Separating reflective and fluorescent components using independent component analysis

the linear sum of the basis function. Dimensions were reduced, and parameters were less than the number of bases. Thus, the solution was obtained as an over-determined system. The input data were obtained using a three-channel RGB color camera under nine types of lights.

Fu et al. [13] decomposed reflections and fluorescence using multispectral lights at high-frequency illuminations in the wavelength domain. For simplicity, we use the example shown in Fig. 8.11 (Fu et al. [13]), where we assume that we can design two types of high-frequency illuminations that are complementary. The shape of the spectrum of one light forms a square wave. The other forms an opposite square wave. For simplicity, we assume that the shape of the spectral absorption is flat in the wavelength domain. Thus, the amount of absorption energies are the same for the two illuminations. Therefore, the shape of fluorescence spectra becomes the same when illuminated by these two lights. However, reflection is bright if the light is bright at a certain wavelength, and it is dark if the light is dark at such a wavelength. As a result, if the brightness of light is zero at a certain wavelength, the observed light of this wavelength consists of only fluorescent light. Additionally, if the brightness of light in a certain wavelength is nonzero, the reflection can be calculated by subtracting the brightness of fluorescence from the input brightness. This explanation shows that simple subtraction can distinguish reflection from fluorescence, and the detailed procedure for actual illumination, which is not an ideal square wave, is explained by Fu et al. [13].

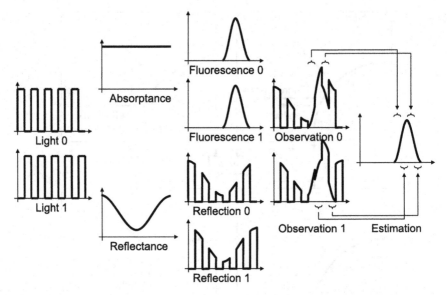

Fig. 8.11 Separating reflective and fluorescent components using high-frequency illumination in the spectral domain

Fig. 8.12 Donaldson matrix representing the absorption and the emittance at each wavelength

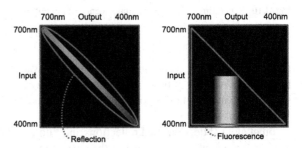

Tominaga et al. [14] explained the method needed to obtain the so-called Donaldson matrix. This matrix (Fig. 8.12) represents the amount of light emitted, depending on the type of light. Row i represents the input wavelength, λ_i, and the column j represents the output wavelength, λ_j. The elements of matrix $d_{i,j}$ represents the brightness emitted from the target material. This value represents the emitted light of wavelength λ_j when the light of wavelength λ_i is illuminated at the material. Each component of the Donaldson matrix includes both reflection and fluorescence. For a material without fluorescence, the diagonal elements may become nonzero, whereas the remaining elements become zero. For a material with fluorescence, the lower triangular elements become nonzero, whereas the upper triangular elements becomes zero. They measured the Donaldson matrix from a multispectral image obtained by a camera where a liquid-crystal-tunable filter was placed in front of it. Because the Donaldson matrix differs, depending on the material, the Donaldson matrix was useful in this case.

8.3 Light Separation

Photometric estimation analyzes the photometric properties of an object's surface. For simplification, it is usually assumed that an object is illuminated by a single light source. Using multiple light sources, however, is desirable in many cases, because it has several advantages. For example, it reduces the time needed for observation and improves SNR. To realize photometric estimation with illumination by multiple light sources, it is necessary to separate their effects and generate images as if the object is illuminated separately by each light source.

Objects are illuminated from multiple directions for photometric stereo, BRDF analysis, and other techniques. Because it is necessary to obtain the effect of each light source, a simple method turns on each light source sequentially. Although controlled illumination is assumed in many active-lighting situations, external lights that are not controllable may affect observations. In such cases, it is impossible to distinguish the system lights from the external lights via simple illumination. In this section, a method to separate the effect of multiple light sources from the observed images is presented.

Figure 8.13 shows a classification method for extracting each light source from multiple ones. If a system emits light and uses a camera to receive it, the problem is regarded as one of a communication system between the light sources and the camera. In communication technology, sending signals from multiple transmitters when using a channel is called multiplexing. For light, the time, wavelength and intensity are often used as the features of light for multiplexing. In Fig. 8.13a, lights from different sources are assigned to different time slots (i.e., time-division multiplexing). It is easy to obtain which light source illuminates an object at a frame if the light source and camera are synchronized. In Fig. 8.13b, if the sensor can detect different wavelengths of light, say, from an RGB camera, it becomes possible to separate them by emitting different wavelengths from multiple sources. In Fig. 8.13c, all light sources emit light simultaneously and are observed in the same camera frame. The sources share

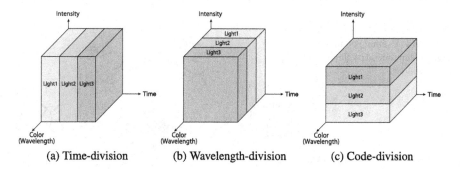

(a) Time-division (b) Wavelength-division (c) Code-division

Fig. 8.13 Examples of methods used to extract each light source from multiple light sources. **a** The timing of light sources are separated. **b** The light sources use different wavelengths. **c** All light sources share the capacity of the sensor

the same capacity as the sensor. To enable extraction of all light source information from the observed intensity, the lights are encoded using multiple camera frames. This approach is called code-division multiplexing and is described next.

8.3.1 Code-Division Multiplexed Illumination

The methods of code-division multiplexed illumination decompose an image of a scene that is simultaneously illuminated by multiple light sources to multiple images of the scene illuminated by each single light source. Schechner et al. [15] proposed a new Hadamard-based multiplexing scheme that projected multiple lights using each pixel of a video projector. The scheme improved SNR for spectroscopy with its multiple acquisition capability of object illumination [16]. Mukaigawa et al. [17] proposed a similar method of estimating the BRDF of an object's surface by illuminating it from various directions.

Let us consider the situation shown in Fig. 8.14, where an object is illuminated by three light sources. If the lights illuminate the object via time-division multiplexing, the relationship between the measurements, a_i, of three frames, $i = 1, 2$, and 3, and the intensity, l_j, at a pixel of the image by the light source, $j = 1, 2$, and 3, is given by

$$\begin{bmatrix} a_1 \\ a_2 \\ a_3 \end{bmatrix} = \begin{bmatrix} 1 & 0 & 0 \\ 0 & 1 & 0 \\ 0 & 0 & 1 \end{bmatrix} \begin{bmatrix} l_1 \\ l_2 \\ l_3 \end{bmatrix}, \tag{8.10}$$

The matrix reveals which light is used on to acquire an image. If the measurements have noise of variance, σ^2, the intensity of a light source has the same variance. If the lights illuminate the object based on code-division multiplexing, an example of coding is that two of three light sources illuminate simultaneously. The relationship is given by

Fig. 8.14 Example of an object illuminated by three light sources

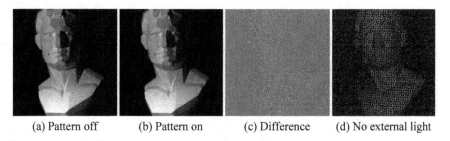

(a) Pattern off   (b) Pattern on   (c) Difference   (d) No external light

Fig. 8.15 The two images on the left are captured while the object is illuminated by the lamp. **a** The laser pattern is turned off, and **b** it is turned on. **c** Their difference. **d** Captured without the lamp

$$\begin{bmatrix} a_1 \\ a_2 \\ a_3 \end{bmatrix} = \begin{bmatrix} 1 & 1 & 0 \\ 0 & 1 & 1 \\ 1 & 0 & 1 \end{bmatrix} \begin{bmatrix} l_1 \\ l_2 \\ l_3 \end{bmatrix}. \tag{8.11}$$

The intensity by each light source is estimated by demultiplexing:

$$\begin{bmatrix} l_1 \\ l_2 \\ l_3 \end{bmatrix} = \frac{1}{2} \begin{bmatrix} 1 & -1 & 1 \\ 1 & 1 & -1 \\ -1 & 1 & 1 \end{bmatrix} \cdot \begin{bmatrix} a_1 \\ a_2 \\ a_3 \end{bmatrix}, \tag{8.12}$$

The variance of the estimated intensity is reduced to $(3/4)\sigma^2$ [15], which is an advantage of code-division-multiplexing illumination. The variance can be minimized by using the Hadamard code, and the SNR ratio is $\sqrt{L}/2$ better than the time-division if L images are used.

This feature can be used for separating the light in a system from the noise from external light not controlled in the system. Even if the external light is very strong (e.g., sunlight), the image illuminated by the weak light source can be extracted using a method based on code-division multiplexing [18]. Figure 8.15 shows an example of an object being illuminated by two light sources. One is a structured-light grid pattern, and the other is a strong lamp. The two images on the left are captured while the object is illuminated by the lamp. Although the grid pattern is turned on in the second image from the left, it is difficult to recognize the pattern. The third image is their difference. The image on the right is captured without the lamp. When the grid pattern is weak, the captured image is noisy, even without the external light.

Code-division multiplexing is applied to this situation to extract the pattern by using codes of length $L = 7, 15, 31, 63, 127$, and 255. Figure 8.16 shows the results of demultiplexing. The leftmost image is the reference image for those that are captured with a long exposure time without external light. The other images are the results of demultiplexing with $L = 15, 63$, and 255. Although the pattern of the input images is almost impossible to recognize, the proposed method succeeded in demultiplexing the pattern. The demultiplexed images are evaluated by comparison

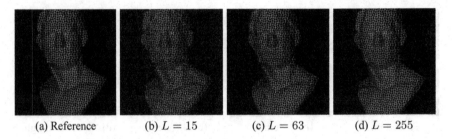

(a) Reference   (b) $L = 15$ (c) $L = 63$ (d) $L = 255$

Fig. 8.16 The results of demultiplexing. **a** The reference image with which to compare images that are captured with long exposure time without external light. The other images are the results with different length of the codes. **b** $L = 15$. **c** $L = 63$. **d** $L = 255$

Fig. 8.17 Peak SNR (PSNR) of demultiplexed images of the wave-grid pattern are evaluated by comparing the reference image

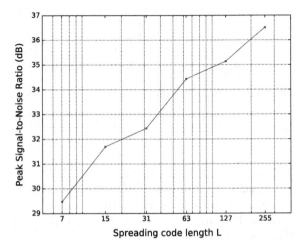

to the reference image. The PSNR for the results is shown in Fig. 8.17 and is improved according to the length of the code.

8.3.2 Applying Code-Division Multiplexed Illumination for a Dynamic Scene

Extending code-division multiplexing to dynamic scenes is beneficial to obtaining scene information. Wenger et al. [19] proposed a method for relighting images of moving persons using video captured via Hadamard-based multiplexed illumination. The method introduced motion compensation by calculating optical flow to demultiplex lights projected onto a moving person in the video.

Another approach was based on signal processing, which observed moving objects without tracking them. Sagawa and Satoh [18] proposed a method that combined multiplexing and high-pass filtering. If the multiplexing code were to be observed at a sufficiently high frame rate, the change in brightness caused by the motion of an object would be slow. Thus, it could be assumed that the motion only affected the low-frequency component. Therefore, the proposed method applied a high-pass filter to the received signal and demultiplexed it after removing the low-frequency component.

A high-pass filter that passes signals with a frequency higher than ω_T (cycles/frame) is expressed in the frequency domain as

$$H(\omega) = \begin{cases} 1 & \omega > \omega_T, \\ 0 & \text{otherwise.} \end{cases} \tag{8.13}$$

The filter, $h(t)$, in the time domain is defined as follows.

$$h(t) = w(t) \circ \text{IDFT}(H(\omega)), \tag{8.14}$$

where IDFT(H) is the inverse discrete Fourier transform of $H(\omega)$, and $w(t)$ is the Hanning-window function defined by $w(t) = 0.5 - 0.5\cos(2\pi t/L_w)$, where L_w is the window length.

Let $\mathbf{s} = [S(t)]$ be the vector that represents the code. By using the row-vector, $\mathbf{h}$, that consists of the high-pass filter values, the $L \times (L + L_w - 1)$ matrix, $\mathbf{H}$, is defined as follows:

$$\mathbf{H} = \begin{bmatrix} \mathbf{h} & 0 & \dots & 0 & 0 \\ 0 & \mathbf{h} & 0 & \dots & 0 \\ & & \dots & & \\ 0 & \dots & 0 & 0 & \mathbf{h}. \end{bmatrix} \tag{8.15}$$

By removing the nonzero low-frequency component of noise by using the high-pass filter, the following equation holds:

$$\mathbf{H}s l = \mathbf{H}\mathbf{a} \tag{8.16}$$

The least-squares solution of this equation is given as

$$l = (\mathbf{s}^\top \mathbf{H}^\top \mathbf{H}\mathbf{s})^{-1}\mathbf{s}^\top \mathbf{H}^\top \mathbf{H}\mathbf{a}. \tag{8.17}$$

The coefficient on the right-hand side of the equation corresponds to the composite function of demultiplexing, which is used as a filter to convolve with the received signal for demultiplexing. Figure 8.18 shows an example of the filter function generated from the coefficient in the case of $L = 31$, $L_w = 8$, and $\omega_T = 1/L_w$. The high-pass filter removes the low-frequency component from the coefficients without the need for a high-pass filter.

In the case of multiple light sources, the individual signal is extracted by using circular-shifted code. Let $\mathbf{s}_j$ be the column vector of the code obtained by circular-

Fig. 8.18 Example of the filter function generated from the coefficient of Eq. (8.17) in the case of $L = 31$, $L_w = 8$, and $\omega_T = 1/L_w$. The high-pass filter removes the low-frequency component from the coefficients without a high-pass filter

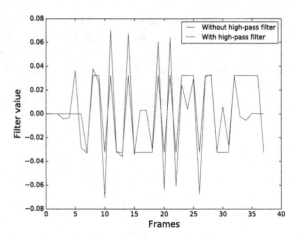

shifting **s** for j times. If M lights are used, the coefficient matrix is obtained as follows, similar to the case of a single light:

$$(\mathbf{Q}^\top \mathbf{H}^\top \mathbf{H} \mathbf{Q})^{-1} \mathbf{Q}^\top \mathbf{H}^\top \mathbf{H}, \tag{8.18}$$

where **Q** is the matrix of the code:

$$\mathbf{Q} = \begin{bmatrix} \mathbf{s}_{j_1} & \mathbf{s}_{j_2} & \cdots & \mathbf{s}_{j_M} \end{bmatrix}. \tag{8.19}$$

Figure 8.19 shows images in which a bouncing ball is observed by projecting a wave-grid pattern for 3D reconstruction under sunlight. The luminance of the sunlight is approximately 50 Klx. The images in row (a) show three moments of a bounce. The images in row (b) are the input images when the laser is turned on. Although the laser power is 85 mW in this experiment, it is nearly impossible to recognize the pattern. The length of the spreading code is $L = 255$. The results of demultiplexing without filtering are shown in row (c). Artifacts occur around the boundary of the ball and hand caused by the motion. The results with temporal filtering are shown in row (d). The artifacts caused by motion are successfully removed from the demultiplexed images. The 3D reconstruction in row (e) is generated for each demultiplexed image by the method of [20].

8.4 Light and Reflectance Decomposition

Image-based object modeling can be used to compose photorealistic images of modeled objects for various rendering conditions, such as viewpoint, light directions, etc. However, it is challenging to acquire the large number of object images required for all combinations of capturing parameters and to then handle the resulting huge data

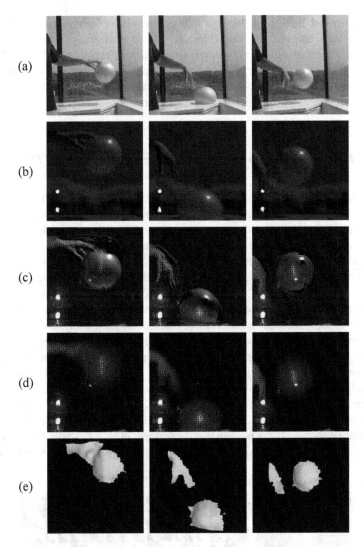

Fig. 8.19 The results of demultiplexing of a bouncing ball under sunlight. The images in row (**a**) show three moments at which the ball bounces. The images in row (**b**) are the input images when the laser is turned on. The results of demultiplexing without filtering are shown in row (**c**). The results with temporal filtering are shown in row (**d**). The 3D reconstruction in row (**e**) is generated for each demultiplexed image by the method of [20]

sets for the model. By using active-lighting systems, objects' geometrical information as well as their complete 4D indexed texture sets, or bi-directional texture functions (BTF) can be obtained in a highly automated manner as explained in Sect. 3.3.5. By using the 4D texture database, light and reflectance information can be decomposed into compact representation. Such decomposed texture database facilitates rendering objects with arbitrary viewpoints, illumination, and deformation.

8.4.1 Decomposition by Tensor Products Expansion (TPE)

Decomposition of 4D texture has been conducted intensively for the purpose of compression of surface light field. One well-known method to compress texture images uses PCA/SVD-based compression [21, 22]. In that research, texture data is rearranged into 2D indexed elements, or a matrix. The matrix is approximated as a sum of the outer products of vectors using SVD. The approximation is more efficient if the row vectors (or column vectors) of the matrix have strong correlations. In the eigentexture method [21], texture images for each polygon are rasterized into vectors specified by a 1D view index. Matrices are formed by aligning these vectors according to view indices. In the work of Chen et al. [22], compression is done in a way similar to the eigentexture method, except textures are first re-sampled for uniformly distributed 2D view directions and then compressed. The re-sampling of textures prevents uneven approximation, and 2D indexing of the view enables efficient hardware-accelerated rendering.

4D texutre set for a single polygon is shown in Fig. 8.20, in which each column of textures corresponds to a certain view direction, and each row corresponds to a certain light direction. It can be seen that the textures in each row tend to have similar average brightness values. This is because diffuse reflection, which amounts to most

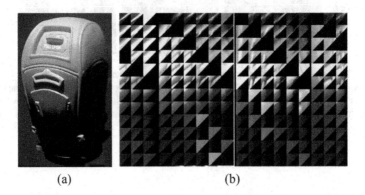

(a)   (b)

Fig. 8.20 Model and texture database. **a** An original image of the modeled object. **b** Visualization of parameterized textures, in which each row of textures is captured from a certain view direction and each column of textures is captured for a certain light direction

of the reflection, tends to depend only on light direction. For each column, texture patterns are most similar because they are captured from fixed view directions. Thus, the changes in brightness and texture pattern are strongly correlated between columns and rows of textures. If the textures are decomposed by PCA/SVD-based techniques such as the eigentexture method, and arranged by the coefficients of eigentextures (principal components) in terms of view and light indices, it is expected that there still remain strong correlations along these indices. To utilize these correlations for efficient compression, the texture database can be decomposed into tensors by using tensor product expansion (TPE).

Tensor and TPE are generalizations of matrices and SVD. As matrices are expanded into sums of products of vectors, tensors can be expanded into sums of tensor products. Let $\mathbf{A}$ be a 3D tensor of size $L \times M \times N$. $\mathbf{A}$ can be expressed as

$$\mathbf{A} = \sum_r \alpha_r \mathbf{u}_r \otimes \mathbf{v}_r \otimes \mathbf{w}_r, \tag{8.20}$$

where r is an index of terms, α_r is a coefficient of term r, $\mathbf{u}_r$, $\mathbf{v}_r$, and $\mathbf{w}_r$ are L-D, M-D, and N-D unit vectors, and the operator $\otimes$ means tensor product. Thus, the form above means

$$A_{i,j,k} = \sum_r \alpha_r u_{r,i} v_{r,j} w_{r,k}. \tag{8.21}$$

$$|\mathbf{u}_r| = 1, |\mathbf{v}_r| = 1, |\mathbf{w}_r| = 1,$$

$$\alpha_r \geq \alpha_s \quad \text{if} \quad r < s$$

where $A_{i,j,k}$ is an element of tensor $\mathbf{A}$ with indices of i, j, k, $u_{r,i}$ is ith element of vector $\mathbf{u}_r$. We can approximate tensor $\mathbf{A}$ by neglecting terms with small significance (i.e., terms with small α_r). Truncating the form into a sum of K terms, we achieve a compression rate of $K(L + M + N)/LMN$.

There are several different ways to pack texture set information into tensors. One of them is to pack a tensor set from a polygon $\{T(i_p, *, *, *, *)\}$ (here, symbols "*" mean "don't care") into a 3D tensor, using the first tensor index for indicating texel, the second for view direction, the third for light direction. This is done by constructing tensors $\mathbf{A}(i_p)$ of size $N_t \times (N_{v\theta} N_{v\phi}) \times (N_{l\theta} N_{l\phi})$ (N_t is the number of texels in a texture) for each polygon $P(i_p)$ by

$$\{\mathbf{A}(i_p)\}_{i, (i_{v\theta} N_{v\phi} + i_{v\phi}), (i_{l\theta} N_{l\phi} + i_{l\phi})}$$
$$= \text{Texel}(i, T(i_p, i_{v\theta}, i_{v\phi}, i_{l\theta}, i_{l\phi})), \tag{8.22}$$

where $\{A(i_p)\}_{i,j,k}$ is an element of tensor $\mathbf{A}(i_p)$ with indices of i, j, k, and $\text{Texel}(i, \cdot)$ denotes ith texel value of a texture. 2D arrangement of textures by view and light indices (i and j in the form above) is the same as the one shown in Fig. 8.20b.

One drawback of this packing is that textures which have strong specular reflection do not align into columns in the arrangement of Fig. 8.20b. Examining the figure, we can see some bright textures aligned in a diagonal direction. Those bright textures include strong specular components. There are also some blank textures aligned in the same way. These textures cannot be captured because the lights for the halogen lamps on the capturing platform are occluded by the circle equipped with cameras for the view/light condition. These diagonally aligned elements are difficult to approximate by the form (8.21), and we have found them to be harmful for TPE compression. Since these textures are almost uniform textures, we subtract DC components from all the textures and approximate only AC components by TPE. DC components are stored separately.

As opposed to SVD, for which there exists a robust algorithm to calculate optimal solution, an algorithm to obtain the optimal solution for TPE is still an open area of research. Murakami et. al proposed a fast calculation method for TPE, applying the power method which was originally used for calculating SVD. Although their algorithm is not guaranteed to produce the optimal solution, we use this method because it is fast and its solution is sufficient for the purpose of compression. A brief description of their algorithm to calculate the expansion of a 3D tensor $\mathbf{A}$ is as follows:

- By iterating the following procedure, obtain $\alpha_s, \mathbf{u}_s, \mathbf{v}_s, \mathbf{w}_s$ of Eq. (8.20) for $s = 1, 2, \ldots$.

  - Initialize $\mathbf{u}_s, \mathbf{v}_s, \mathbf{w}_s$ as arbitrary unit vectors.
  - Obtain the residual tensor $\mathbf{R}$ through the operation $\mathbf{R} \leftarrow \mathbf{A} - \sum_{r=1}^{s-1} \alpha_s \mathbf{u}_s \otimes \mathbf{v}_s \otimes \mathbf{w}_s$.
  - If $(\|\mathbf{R}\|/\|\mathbf{A}\|)^2$ is less than ϵ, stop the calculation, where $\|\cdot\|$ means the 2-norm of a tensor (the root of the sum of squared elements in a tensor), and ϵ is the tolerable squared error rate.
  - Iteratively update $\mathbf{u}_r, \mathbf{v}_r, \mathbf{w}_r$ until these vectors converge by applying the following steps.

    Obtain $\tilde{\mathbf{u}}_r, \tilde{\mathbf{v}}_r, \tilde{\mathbf{w}}_r$ by the following tensor contraction:

    $$\tilde{u}_{r,i} \leftarrow \sum_{j=1}^{M} \sum_{k=1}^{N} R_{i,j,k}\, v_{r,j}\, w_{r,k},$$

    $$\tilde{v}_{r,j} \leftarrow \sum_{k=1}^{N} \sum_{i=1}^{L} R_{i,j,k}\, w_{r,k}\, u_{r,i},$$

    $$\tilde{w}_{r,k} \leftarrow \sum_{i=1}^{L} \sum_{j=1}^{M} R_{i,j,k}\, u_{r,i}\, v_{r,j},$$

    Update $\mathbf{u}_r, \mathbf{v}_r, \mathbf{w}_r$ as the normalized $\tilde{\mathbf{u}}_r, \tilde{\mathbf{v}}_r, \tilde{\mathbf{w}}_r$: $\mathbf{u}_r \leftarrow \frac{\tilde{\mathbf{u}}_r}{|\tilde{\mathbf{u}}_r|}, \mathbf{v}_r \leftarrow \frac{\tilde{\mathbf{v}}_r}{|\tilde{\mathbf{v}}_r|}, \mathbf{w}_r \leftarrow \frac{\tilde{\mathbf{w}}_r}{|\tilde{\mathbf{w}}_r|}$.

To test the efficiency of the compression method, we compressed textures using an SVD technique (eigentexture method) and TPE-based compression with two different dimensions. The sample object was a toy post shown in Fig. 8.20. Each texture has 136 pixels. There exist 2592 ($12 \times 3 \times 12 \times 6$) textures for each polygon. Intervals of azimuth and elevation angles are $30°$. For SVD, we packed the pixel values into matrices $\mathbf{B}(i_p)$ with a size of 136×2592, which can be expressed as

$$\{\mathbf{B}(i_p)\}_{i,\,(i_{v\theta} N_{v\phi} N_{l\theta} N_{l\phi}+i_{v\phi} N_{l\theta} N_{l\phi}+i_{l\theta} N_{l\phi}+i_{l\phi})}$$
$$= \mathrm{Texel}(i, T(i_p, i_{v\theta}, i_{v\phi}, i_{l\theta}, i_{l\phi})).$$

To apply TPE, two methods are possible. One method consists of packing textures into 3D-tensors $\mathbf{A}$ whose size is $136 \times 36 \times 72$, where the three tensor indices correspond to texel location, view direction ($i_{v\theta}$ and $i_{v\phi}$), and light direction ($i_{l\theta}$ and $i_{l\phi}$), respectively, using the form (8.22). The other method is packing textures into 4D-tensors $\mathbf{C}$ whose size is $136 \times 36 \times 12 \times 6$, where the four tensor indices correspond to texel location, view direction ($i_{v\theta}$ and $i_{v\phi}$), and light direction indices ($i_{l\theta}$ and $i_{l\phi}$). The packing is done by the form

$$\{\mathbf{C}(i_p)\}_{i,\,(i_{v\theta}N_{v\phi}+i_{v\phi}),i_{l\theta},i_{l\phi}}$$
$$= \mathrm{Texel}(i, T(i_p, i_{v\theta}, i_{v\phi}, i_{l\theta}, i_{l\phi})).$$

Table 8.1 shows an example of decomposition, which describes data sizes needed to store each term, average numbers of terms needed to approximate textures for each polygon, and average data sizes for each polygon including stored AC components. It was assumed that the compressed data were expressed as a collection of 2 byte short numbers. Because the freedom of the approximation model decreases in order of SVD, 3D TPE, and 4D TPE, the number of terms needed for approximation increases in the same order. TPE-based compression uses less size to store 1 term of the expanded data, but it needs more terms for approximation. As a result, data size 3D TPE compression was about 2.4 times less than SVD. Although the data size for 1 term of 4D TPE was smaller than that of 3D TPE, the average data size of 4D TPE compression was larger than 3D TPE because of the increased number of terms. Figure 8.21 plots the data sizes of compressed textures for each polygon compressed using SVD and 3D TPE method. The horizontal axis represents polygon index i_p, and the vertical represents the compressed size of the texture data.

Table 8.1 Compression result

	Data size (1 term)	Average number of term	Average data size
SVD	5456	8.56	46703
3D TPE	488	23.22	19107
4D TPE	380	34.99	21072

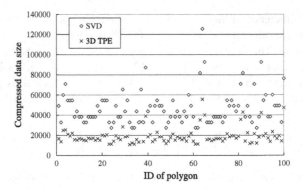

Fig. 8.21 Data sizes of compressed textures (SVD and TPE)

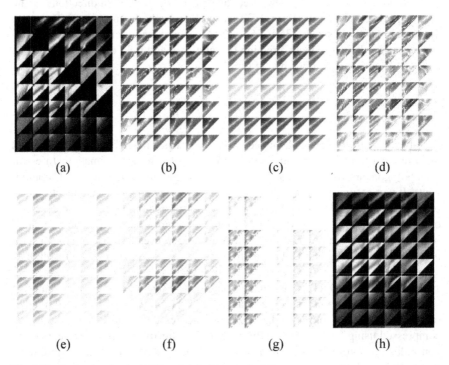

Fig. 8.22 Approximation by TPE. **a** Original image. **b** AC components of textures. **c** Term 1 of TPE. **d** Residual image after subtraction of term 1. **e** Term 3. **f** Term 5. **g** Term 9. **h** Result image

Figure 8.22 shows how TPE approximates textures. Figure 8.22b is the AC components of the original texture (a). (b) is approximated by term 1 (c) and the residual of term 1 is (d). The texture is approximated by terms including term 3, 5, and 9 (shown in (e)–(g)). The resulting approximation is shown in (h).

8.4.2 Synthesis Under Arbitrary View and Light Positions

By using the tensor texture, the texture under arbitrary view and light positions can be synthesized. Let us assume that we have vertices $V(i_v)(0 \le i_v < N_v)$ which form polygons $P(i_p)(0 \le i_p < N_p)$, where i_v represents the index of vertices, $V(i_v)$ is the vertex specified by index i_v and N_v is the number of vertices. Polygon $P(i_p)$ consists of three vertices $V(t(i_p, j))(0 \le j \le 2)$, where $t(i_p, j)$ is a table that enables us to look up vertex indices by polygon indices.

For the rendering process, the user can specify virtual camera position, light direction, object position (translation and rotation), and geometrical deformation. From the camera position and object position, rotation from the camera coordinate system to the object coordinate system, which is expressed as R_o, is calculated. Let us assume that view direction and light direction can be expressed by $\mathbf{v}_c$ and $\mathbf{l}_c$ in the camera coordinate system. Normally, $\mathbf{v}_c$ is a fixed vector (for example $[0, 0, -1]^t$). Thus, view and light directions expressed by the object coordinate system are $R_o \mathbf{v}_c$, $R_o \mathbf{l}_c$.

From the given deformation, the 3D rotation of surface point $V(i_v)$ relative to the object coordinate system, which can be expressed as $R_d(i_v)$, can be calculated. If there is no geometrical deformation, $R_d(i_v) = I$ (I is an identity rotation) for all vertices. Relative rotation of each vertex $V(i_v)$ from the camera coordinate system can be expressed as $R_d(i_v) \circ R_o$.

$R_d(i_v)$ may be directly calculated if a mathematical model of the deformation is given. If it is not available, $R_d(i_v)$ from a geometrical transformation caused by deformation can be can calculated. To do so, the normal vector at $V(i_v)$ with and without deformation, which we describe as $\mathbf{n}'(i_v)$ and $\mathbf{n}(i_v)$, are calculated. Further, normalized direction vectors of edges connected to vertex $V(i_v)$ with and without deformation, which described as $\mathbf{e}'(i_v, j)(j = 0, 1, 2, \ldots, E(i_v) - 1)$ $\mathbf{e}(i_v, j)(j = 0, 1, 2, \ldots, E(i_v) - 1)$, where $E(i_v)$ denotes the number of edges connected to vertex $V(i_v)$, can be calculated. Then the rotation $R_d(i_v)$ such that $\mathbf{n}'(i_v) \approx R_d(i_v)\mathbf{n}(i_v)$, $\mathbf{e}'(i_v, j) \approx R_d(i_v)\mathbf{e}(i_v, j)(j = 0, 1, 2, \ldots, E(i_v) - 1)$ is obtained. For the calculation, the method proposed by Arun et al. [23] can be used. Now, we get

$$\mathbf{v}_r(i_v) \equiv (R_d(i_v) \circ R_o)\mathbf{v}_c$$

$$\mathbf{l}_r(i_v) \equiv (R_d(i_v) \circ R_o)\mathbf{l}_c,$$

where $\mathbf{v}_r(i_v)$ and $\mathbf{l}_r(i_v)$ are view and light directions at vertex $V(i_v)$ expressed by object coordinate system. Describing azimuth and elevation angles of the direction vector by $\text{azm}(\cdot)$ and $\text{elv}(\cdot)$, four angles $\text{azm}(\mathbf{v}_r(i_v))$, $\text{elv}(\mathbf{v}_r(i_v))$, $\text{azm}(\mathbf{l}_r(i_v))$, and $\text{elv}(\mathbf{l}_r(i_v))$ have direct correspondence to indices of the texture database, $i_{v\theta}$, $i_{v\phi}$, $i_{l\theta}$, and $i_{l\phi}$. Since the pair of view and light directions $[\mathbf{v}_r(i_v)), \mathbf{l}_r(i_v)]$ represent conditions of the texture used for rendering, it is called "rendering condition." Also, the view direction of the pair is called "rendering view condition," and the light direction "rendering light condition."

Textures in the BTF database are sampled at discrete directions of view and light, so the sample textures must be interpolated to generate textures needed for rendering. We call the view/light direction pairs in the database "sample conditions," and we use the terms "sample view conditions" and "sample light conditions" in a similar way as rendering conditions. If sample view conditions or sample light conditions are plotted regarding their azimuth and elevation angles as 2D orthogonal coordinates, the plots form lattice points aligned by fixed intervals for each axis.

Then, textures corresponding to calculated rendering conditions $[\mathbf{v}_r(i_v), \mathbf{l}_r(i_v)]$ are generated by using the weighted sum of neighbor samples. Let $\tilde{\mathbf{v}}_0(i_v)$, $\tilde{\mathbf{v}}_1(i_v)$, $\tilde{\mathbf{v}}_2(i_v)$ be three neighbor sample view conditions of a rendering view condition $\mathbf{v}_r(i_v)$. In addition, let $W_0^v(i_v)$, $W_1^v(i_v)$, $W_2^v(i_v)$ be weights for the neighbor sample view conditions, fulfilling the constraint $W_0^v(i_v) + W_1^v(i_v) + W_2^v(i_v) = 1$. Selection of the neighbor sample view conditions is done by the following process: Let us define

$$i_{v\theta}^- \equiv \left\lfloor \frac{\mathrm{azm}(\mathbf{v}_r(i_v))}{\Delta_{v\theta}} \right\rfloor,$$

$$i_{v\phi}^- \equiv \left\lfloor \frac{\mathrm{elv}(\mathbf{v}_r(i_v))}{\Delta_{v\phi}} \right\rfloor,$$

$$r_{v\theta} \equiv azm(\mathbf{v}_r(i_v)) - i_{v\theta}^- \Delta_{v\theta},$$

$$r_{v\phi} \equiv \mathrm{elv}(\mathbf{v}_r(i_v)) - i_{v\phi}^- \Delta_{v\phi},$$

where $\lfloor \cdot \rfloor$ denotes floor function. Since $\Delta_{v\theta}$ and $\Delta_{v\phi}$ are intervals for azimuth and elevation angle of sample view conditions, $(azm(\mathbf{v}_r(i_v)), \mathrm{elv}(\mathbf{v}_r(i_v)))$ exists in the region surrounded by $(i_{v\theta}^- \Delta_{v\theta}, i_{v\phi}^- \Delta_{v\phi})$, $((i_{v\theta}^- + 1)\Delta_{v\theta}, i_{v\phi}^- \Delta_{v\phi})$, $(i_{v\theta}^- \Delta_{v\theta}, (i_{v\theta}^- + 1)\Delta_{v\phi})$ and $((i_{v\theta}^- + 1)\Delta_{v\theta}, (i_{v\phi}^- + 1)\Delta_{v\phi})$. Then sample view conditions and their weights are defined as

$$\begin{bmatrix} \mathrm{azm}(\tilde{\mathbf{v}}_0(i_v)) \\ \mathrm{elv}(\tilde{\mathbf{v}}_0(i_v)) \end{bmatrix} \equiv \begin{cases} \begin{bmatrix} i_{v\theta}^- \Delta_{v\theta} \\ i_{v\phi}^- \Delta_{v\phi} \end{bmatrix} & \text{if } (r_{v\theta} + r_{v\phi}) \leq 1 \\ \begin{bmatrix} (i_{v\theta}^- + 1)\Delta_{v\theta} \\ (i_{v\phi}^- + 1)\Delta_{v\phi} \end{bmatrix} & \text{otherwise,} \end{cases}$$

$$\begin{bmatrix} \mathrm{azm}(\tilde{\mathbf{v}}_1(i_v)) \\ \mathrm{elv}(\tilde{\mathbf{v}}_1(i_v)) \end{bmatrix} \equiv \begin{bmatrix} (i_{v\theta}^- + 1)\Delta_{v\theta} \\ i_{v\phi}^- \Delta_{v\phi} \end{bmatrix},$$

$$\begin{bmatrix} \mathrm{azm}(\tilde{\mathbf{v}}_2(i_v)) \\ \mathrm{elv}(\tilde{\mathbf{v}}_2(i_v)) \end{bmatrix} \equiv \begin{bmatrix} i_{v\theta}^- \Delta_{v\theta} \\ (i_{v\phi}^- + 1)\Delta_{v\phi} \end{bmatrix},$$

$$\begin{bmatrix} W_0^v(i_v) \\ W_1^v(i_v) \\ W_2^v(i_v) \end{bmatrix} \equiv \begin{cases} \begin{bmatrix} 1 - (r_{v\theta} + r_{v\phi}) \\ r_{v\theta} \\ r_{v\phi} \end{bmatrix} & \text{if}(r_{v\theta} + r_{v\phi}) \leq 1 \\ \begin{bmatrix} (r_{v\theta} + r_{v\phi}) - 1 \\ 1 - r_{v\phi} \\ 1 - r_{v\theta} \end{bmatrix} & \text{otherwise.} \end{cases}$$

By the definitions above, three sample view conditions $\tilde{\mathbf{v}}_m(i_v)(m = 0, 1, 2)$ are selected so that the triangle they form includes the rendering view condition $\mathbf{v}_r(i_v)$ in the orthogonal coordinate plane of azimuth and elevation angles, and we can regard the triple of weights $[W_0^v(i_v), W_1^v(i_v), W_2^v(i_v)]^t$ as barycentric coordinates for the view condition in the azimuth-elevation coordinate space. If the rendering view condition $\mathbf{v}_r(i_v)$ is placed on the sample view condition $\tilde{\mathbf{v}}_0(i_v)$, the weight $W_0^v(i_v)$ is 1 and linearly decreases to 0 as $\mathbf{v}_r(i_v)$ moves toward the opposite side of the triangle formed by the three sample view conditions.

For the light direction, let $\tilde{\mathbf{l}}_0(i_v), \tilde{\mathbf{l}}_1(i_v), \tilde{\mathbf{l}}_2(i_v)$ be the three neighbor sample light conditions of the rendering light condition $\mathbf{l}_r(i_v)$, and let $W_0^l(i_v), W_1^l(i_v)$, and $W_2^l(i_v)$ under constraint $W_0^l(i_v) + W_1^l(i_v) + W_2^l(i_v) = 1$ be the weights for the sample light conditions. Weights can be seen as barycentric coordinates for the light condition. Neighbor sample light conditions and their weights are calculated in similar way as that of sample view conditions and their weights which is described above.

Using the above notations, we can generate texture T_g of polygon $P(i_p)$ calculated from the rendering condition on vertex $V(i_v)$ as

$$T_v(i_p, \mathbf{v}_r(i_v), \mathbf{l}_r(i_v))$$

$$\equiv \sum_{m=0}^{2} \sum_{n=0}^{2} W_m^v(i_v) W_n^l(i_v) T(i_p, \tilde{\mathbf{v}}_m(i_v), \tilde{\mathbf{l}}_n(i_v))$$

$$T(i_p, \tilde{\mathbf{v}}_m(i_v), \tilde{\mathbf{l}}_n(i_v))$$

$$\equiv T\left(i_p, \frac{\mathrm{azm}(\tilde{\mathbf{v}}_m(i_v))}{\Delta_{v_\theta}}, \frac{\mathrm{elv}(\tilde{\mathbf{v}}_m(i_v))}{\Delta_{v_\phi}}, \frac{\mathrm{azm}(\tilde{\mathbf{l}}_n(i_v))}{\Delta_{l_\theta}}, \frac{\mathrm{elv}(\tilde{\mathbf{l}}_n(i_v))}{\Delta_{l_\theta}}\right).$$

Note that $\frac{\mathrm{azm}(\tilde{\mathbf{v}}_m(i_v))}{\Delta_{v_\theta}}, \frac{\mathrm{elv}(\tilde{\mathbf{v}}_m(i_v))}{\Delta_{v_\phi}}, \frac{\mathrm{azm}(\tilde{\mathbf{l}}_m(i_v))}{\Delta_{l_\theta}}$, and $\frac{\mathrm{elv}(\tilde{\mathbf{l}}_m(i_v))}{\Delta_{l_\phi}}$ are all integers for $m = 0, 1, 2$ because $[\tilde{v}_m, \tilde{l}_m]$ are sample conditions where corresponding textures exist in the BTF database.

The final texture $T_p(i_p)$ of polygon $P(i_p)$ used for rendering is generated by blending three textures, and is calculated from the rendering conditions on three vertices forming the polygon, $V(t(i_p, j))(j = 0, 1, 2)$. The blended textures are

$$T_v(i_p, \mathbf{v}_r(t(i_p, m)), \mathbf{l}_r(t(i_p, m)), (m = 0, 1, 2).$$

The purpose of this process is to minimize the texture gap between polygons. This blending is done in the same way that pixel values of three vertices are blended when Gouraud shading is applied. Suppose that the texture coordinates $(0, 0)$, $(1, 0)$, and $(0, 1)$ are mapped to the vertices $V(t(i_p, 0))$, $V(t(i_p, 1))$, and $V(t(i_p, 2))$ respectively, and $(s_0(i), s_1(i))$ denote texture coordinates of ith texel. Then the texture $T_p(i_p)$ can be expressed as

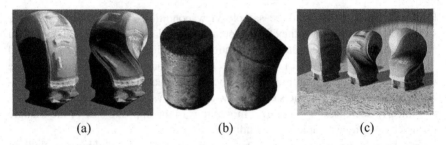

     (a)   (b)   (c)

Fig. 8.23 Rendering results. **a** A toy post. **b** A can wrapped with shiny paper. **c** Synthesis with model based 3D CG system

$$
\begin{aligned}
\text{Texel}(i, T_p(i_p)) =&(1 - s_0(i) - s_1(i)) \, \text{Texel}(i, T_v(i_p, \mathbf{v}_r(t(i_p, 0)), \mathbf{l}_r(t(i_p, 0)))) \\
&+ s_0(i) \, \text{Texel}(i, T_v(i_p, \mathbf{v}_r(t(i_p, 1)), \mathbf{l}_r(t(i_p, 1)))) \\
&+ s_1(i) \, \text{Texel}(i, T_v(i_p, \mathbf{v}_r(t(i_p, 2)), \mathbf{l}_r(t(i_p, 2)))).
\end{aligned} \tag{8.23}
$$

Figure 8.23 shows synthesis results using Fig. 8.20 and a can wrapped with shiny paper (an example which has complicated surface attributes). Synthesized objects are further merged with conventional model based 3D CG software as shown in Fig. 8.23c. The object in the center of the image is the image-based rendered object, while the rest of the scene was rendered with traditional 3D CG software. Although only one illumination is set in this situation, the scene looks natural and the object was rendered photo-realistically.

References

1. Morimoto T, Chiba M, Katayama Y, Ikeuchi K (2009) Multispectral image analysis for bas-relief at the inner gallery of Bayon temple. In: 22nd CIPA symposium, Kyoto, Japan
2. Blasinski H, Farrell J, Wandell B (2017) Designing illuminant spectral power distributions for surface classification. In: Proceedings of the IEEE conference on computer vision and pattern recognition (CVPR), pp 2164–2173
3. Skaff S, Tin SK, Martinello M (2015) Learning optimal incident illumination using spectral bidirectional reflectance distribution function images for material classification. J Imaging Sci Technol 59(6):1–60405
4. Sato M, Yoshida S, Olwal A, Shi B, Hiyama A, Tanikawa T, Hirose M, Raskar R (2015) Spectrans: versatile material classification for interaction with textureless, specular and transparent surfaces. In: Proceedings of the 33rd annual ACM conference on human factors in computing systems. ACM, pp 2191–2200
5. Liu C, Gu J (2013) Discriminative illumination: per-pixel classification of raw materials based on optimal projections of spectral brdf. IEEE Trans Pattern Anal Mach Intell (PAMI) 36(1):86–98
6. Wang C, Okabe T (2017) Joint optimization of coded illumination and grayscale conversion for one-shot raw material classification. In: Proceedings of the British machine vision conference (BMVC)

7. Wolff LB, Boult TE (1991) Constraining object features using a polarization reflectance model. IEEE Trans Pattern Anal Mach Intell (PAMI) 7:635–657
8. Martínez-Domingo MÁ, Valero EM, Hernández-Andrés J, Tominaga S, Horiuchi T, Hirai K (2017) Image processing pipeline for segmentation and material classification based on multispectral high dynamic range polarimetric images. Opt Express 25(24):30073–30090
9. Chen H, Wolff LB (1998) Polarization phase-based method for material classification in computer vision. Int J Comput Vis (IJCV) 28(1):73–83
10. Howell JR, Menguc MP, Siegel R (2015) Thermal radiation heat transfer. CRC Press
11. Zhang C, Sato I (2012) Image-based separation of reflective and fluorescent components using illumination variant and invariant color. IEEE Trans Pattern Anal Mach Intell (PAMI) 35(12):2866–2877
12. Fu Y, Lam A, Sato I, Okabe T, Sato Y (2015) Reflectance and fluorescence spectral recovery via actively lit RGB images. IEEE Trans Pattern Anal Mach Intell (PAMI) 38(7):1313–1326
13. Fu Y, Lam A, Sato I, Okabe T, Sato Y (2013) Separating reflective and fluorescent components using high frequency illumination in the spectral domain. In: Proceedings of the international conference on computer vision (ICCV), pp 457–464
14. Tominaga S, Hirai K, Horiuchi T (2018) Estimation of fluorescent donaldson matrices using a spectral imaging system. Opt Express 26(2):2132–2148
15. Schechner Y, Nayar S, Belhumeur P (2003) A theory of multiplexed illumination. Proc Int Conf Comput Vis (ICCV) 2:808–815
16. Harwit M, Sloane NJ (1979) Hadamard transform optics. Academic Press
17. Mukaigawa Y, Sumino K, Yagi Y (2007) Multiplexed illumination for measuring brdf using an ellipsoidal mirror and a projector. In: Proceedings of the Asian conference on computer vision (ACCV). Springer, pp 246–257
18. Sagawa R, Satoh Y (2017) Illuminant-camera communication to observe moving objects under strong external light by spread spectrum modulation. In: Proceedings of the IEEE conference on computer vision and pattern recognition (CVPR), pp 5097–5105
19. Wenger A, Gardner A, Tchou C, Unger J, Hawkins T, Debevec P (2005) Performance relighting and reflectance transformation with time-multiplexed illumination. ACM Trans Graph (TOG) 24(3):756–764
20. Sagawa R, Sakashita K, Kasuya N, Kawasaki H, Furukawa R, Yagi Y (2012) Grid-based active stereo with single-colored wave pattern for dense one-shot 3D scan. In: 2012 second international conference on 3d imaging, modeling, processing, visualization & transmission. IEEE, pp 363–370
21. Nishino K, Sato Y, Ikeuchi K (1999) Eigen-texture method: appearance compression based on 3D model. In: Proceedings of the IEEE conference on computer vision and pattern recognition (CVPR), vol 1. IEEE, pp 618–624
22. Chen WC, Bouguet JY, Chu MH, Grzeszczuk R (2002) Light field mapping: efficient representation and hardware rendering of surface light fields. ACM Trans Graph (TOG) 21(3):447–456
23. Arun KS, Huang TS, Blostein SD (1987) Least-squares fitting of two 3-D point sets. IEEE Trans Pattern Anal Mach Intell (PAMI) 5:698–700

Part III
Application

Chapter 9
Visualization/AR/VR/MR Systems

As is shown in previous chapters, active lighting is a powerful tool that satisfies various demands. Succeeding chapters provide several application examples of active lighting. Projecting active-lighting images onto real objects augments their appearance. Such projection mapping is now widely used in various fields, including the entertainment industry. Actively lighting multispectral lights onto oil paintings enables novel art modifications. Multispectral light can deceive human eyes because of our RGB limitations, because the light is represented using continuous wavelengths. Therefore, multispectral light can augment the visualization of real objects. Active lighting also enables us to capture the depth of the human body. Thus, a human pose can be estimated from the analysis of its depth. To represent a digital character, capturing actual human motion enables realistic duplication. Estimating human body positions is necessary for representing the motion of digital character and user positions in augmented-, virtual-, and mixed-reality systems.

9.1 Projection Mapping

Projectors are normally used to project images onto a planar white screen. However, some systems project images onto 3D objects to enhance their appearance. Such projection mapping can represent dynamic scenes, even if the target object is static, because the projector can transmit a sequence of images that dynamically change and animate. The projection mapping presented by Raskar et al. [1] (i.e., the Shader Lamp) strongly impacted the field of computer graphics. Compared to the virtual world shown on a display monitor, projection mapping augments the appearance of real objects to seemingly change reality. Many representations can be exhibited this way, hence the value for entertainment purposes increased (Fig. 9.1). One famous

© Springer Nature Switzerland AG 2020
K. Ikeuchi et al., *Active Lighting and Its Application for Computer Vision*,
Advances in Computer Vision and Pattern Recognition,
https://doi.org/10.1007/978-3-030-56577-0_9

Fig. 9.1 Projection
mapping: changing the
appearance of a real 3D
object using a projector

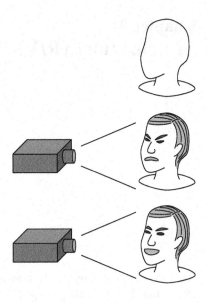

application is the Haunted Mansion experience of Disney theme parks [2]. Moving
pictures of singing ghosts are projected onto a bust that, in reality, lacks facial details.
There are also educational purposes. For example, Bimber et al. [3] used projection
mapping for a fossil exhibit. They projected a skin image of a dinosaur onto its
fossil, so that visitors could see and further imagine the relationships between fossils
and extinct animals. Many firms (e.g., teamLab Inc., Japan) are fully engaged in the
projection mapping business.

9.1.1   BRDF Reproduction on Real Objects

The exhibition of cultural assets has benefited from projection mapping.
Mukaigawa et al. [4] projected a rendered BRDF image of Bizen-Hidasuki onto plaster. The system is shown in Fig. 9.2, and the result is shown in Fig. 9.3. Figure 9.3a
is the actual object that Mukaigawa wanted to reproduce. Figure 9.3b is the plaster
that was used as the projection target. As shown in Fig. 9.3c, projecting the rendered images using Bizen-Hidasuki's BRDF appearance, the plaster's appearance
changed without any application of paint or placement of artifacts. This way, pre-
cious assets can be observed close-up without exposing them to potential harm. In
such a display, the specular reflection from the system remains unchanged unless the
observer's location is tracked. However, the brightness of the actual specular reflection in the real world changes, depending on the observer's position. To overcome
this drawback, Hamasaki et al. [5] presented a diffuse reflection using a projector
while they presented the specular reflection using a head-mounted display (HMD).
Because the HMD tracked the observer's location, the system represented a view-

Fig. 9.2 Projection mapping
for reproducing
Bizen-Hidasuki's appearance

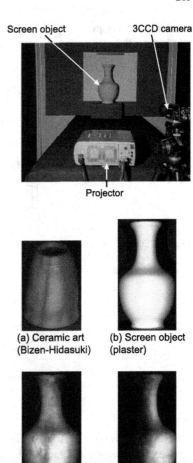

Fig. 9.3 Projection mapping
of Bizen-Hidasuki.
a Bizen-Hidasuki. **b** The
plaster used as a screen.
c Projection mapping result

dependent specular reflection. The observer could see the AR-synthesized image, where the view-dependent specular reflection was presented by the HMD, and the view-independent diffuse reflection was presented by the projector.

9.1.2 Projecting Additional Information to Real Scene

Presenting the information about a working space for employees in factories and warehouses is helpful for several reasons. Imagine a situation in which a warehouse worker needs to know where certain merchandise is stored. The handheld projector

Fig. 9.4 RFIG Lamps:
interacting with a
self-describing world via
photo-sensing wireless tags
and projectors

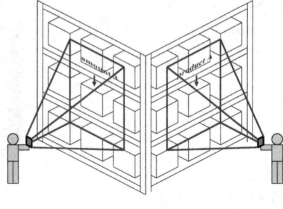

Fig. 9.5 Direction system
for assisting the work in eal
world from a distant site

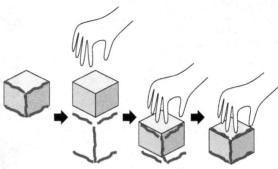

developed by Raskar et al. [6] (Fig. 9.4) can be used to display directions from the
worker to the location. A radio-frequency identification tag is stitched onto each item
so that its approximate position can be determined in 3D space. A photo-sensor is
also attached so that the precise position can be illuminated from the pattern projected
light.

A tele-direction system developed by Hiura et al. [7] (Fig. 9.5) presents worker
instructions to employees at a remote location. First, the projector-camera system
measures the shape of their workspace. Then, the 3D geometrical data is sent from
the instructor to the remote workers.

9.1.3   Hallucinating the Motion of the Object by Projected
Images

Suppose that we want to project an image to another still image mounted on a
wall. Using projection, the still image could be made to look as if it was animated.
Kawabe et al. [8] called this a *deformation-lamp* (hengentou) technique. Intriguingly,
the system developed by Punpongsanon et al. [9] modified the apparent softness of
a curtain as it was actually deformed by the wind. This was extremely innovative in

Fig. 9.6 Haptic projection mapping

that the camera captured the motion of the waving curtain and calculated its optical flow in real time, so that the projected image was properly scaled to exaggerate the dynamics of the curtain, making it appear to have softer material.

9.1.4 Haptic Interaction Realized by Projection Mapping

Projection mapping of heat onto a target can be accomplished using an infrared projector. An experiment performed by Iwai et al. [10] used a halogen lamp instead of an infrared projector. Their system projected a virtual computer-graphics (CG) character onto an examinee's arm using the projector as the infrared beam was focused onto the same place, causing the examinee to "feel" the CG character's existence. Haptic sensitivity can also be simulated using projection mapping [11–13] (Fig. 9.6). For example, if we were to press a finger against a fixed object, a projected image that shows the object giving way or deforming under our touch will cause us to sense that we actually moved or penetrated the object.

9.1.5 Use of 3D Printer in Projection Mapping

When the target screen receiving the projection is not a plane but another solid object, the projected image will obviously be distorted. Recently, 3D-printer performance has drastically improved, and now printers can print 3D objects with transparency. Thus, we can now compensate for the distorted projected image by using the 3D printed object with transparency, while the distorted projected image is corrected by the refraction caused inside the 3D printed object [14].

In case we want to project an image from the projector to 3D object adequately, we must know the geometrical location of the target. We could add markers to the target object for this purpose, but these often obscure the desired result. Recent advances in 3D printing have allowed us to printer objects comprising multiple materials. If we embed the markers inside of an object in this fashion, we can avoid obscuring the desired result. A crucial problem with this approach is that the markers cannot be observed visually by a camera. To overcome the problem, Asayama et al. [15] embedded the markers close to the surface of a semitransparent object. Under visible illumination, the markers were invisible. However, with NIR, they could be seen by the camera.

9.1.6 Projection Mapping as an Innovative Art

The system proposed by Amano [16] projected shading imagery onto a paper (Fig. 9.7). The target paper contained a pseudo-representation of a surface normal. The camera detected the pseudo-color and calculated the shading image, projecting it onto the sheet. The system was robust, because it did not need to recognize the position of the paper, calculate the shape, or reference a database of printed objects.

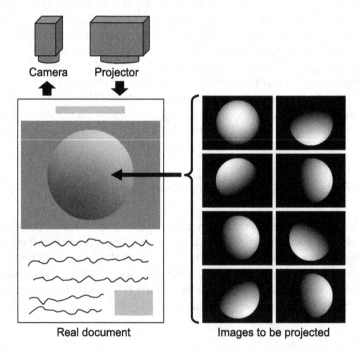

Fig. 9.7 Shading illusion: after recognizing the pseudo-representation of the surface normal printed on actual paper, the rendered shading image is projected

The target sheet was easily created and distributed as printed material, and various shapes were printed. Projecting shading images rendered under different light directions enables observers to view geometrical or bas-relief representations of the printed object. The system can also be used for low-cost advertising, because affected documents can be printed using ordinary printers.

If we project a shading image onto a planar object, we can view the object as if it were a 3D object of the same shading. Feelings of solidity increase if we use stereoscopic glasses. Okutani et al. [17] accomplished this, projecting simulations of carpet, leather, wood, and stone onto the target object. Suppose that we wish to make a 3D product whose surface material will actually be made of leather. With a CAD-like system built upon this framework, a designer can prototype the object with material affinity.

9.1.7 Metamer Enabled by Projecting the Multispectral Light

Multispectral analysis is useful for material estimation. Analysis of multispectral light has a wide variety of applications. For example, actively projecting multispectral light onto oil paintings enables novel re-expressions of art. Multispectral light can deceive our perception because of the aforementioned RGB effect. Thus it is useful for AR purposes.

In Sect. 3.3.1, we introduced the multispectral camera. Some multispectral cameras obtain multispectral images by setting filters or other kinds of optical elements in front of a monochrome camera. Inversely, we can obtain multispectral images from monochrome cameras if we illuminate the scene using different spectral light spectra consecutively. Park et al. [18] made two types of light from the combination of five types of LEDs and observed the scene with an RGB camera under two of the LEDs. They represented the spectral reflectance of the scene as a linear sum of eight bases and recovered the spectral reflectance from the captured two images.

The phenomenon by which two objects are recognized as having different colors under one light source but as the same color under another light source is called *metamerism*. Bala et al. [19] created watermarks using the technique. Cyan-magenta-yellow-key (CMYK) printers can be used to express black-colored prints via a CMY combination or with a single key-black (K) ink. These colors appeared the same under natural light but differently when illuminated with LEDs of certain wavelengths. They selected an LED having a peak wavelength at which the brightness of two inks was sufficiently far apart to be distinguished visually.

We next explain the method of Miyazaki et al. [20] that automatically calculates the mixing ratios for LEDs to generate metamerisms. The observed values of human eyesight, $\mathbf{x} = (X, Y, Z)^{\top}$, can be expressed as follows:

$$\mathbf{x} = \mathbf{PLb} . \tag{9.1}$$

We express the discretized data of the color-matching functions as the $3 \times N_b$ matrix, $\mathbf{P}$, and place the X, Y, and Z color-matching functions in each row:

$$\mathbf{P} = \begin{pmatrix} \bar{x}_1 & \bar{x}_2 & \cdots & \bar{x}_{N_b} \\ \bar{y}_1 & \bar{y}_2 & \cdots & \bar{y}_{N_b} \\ \bar{z}_1 & \bar{z}_2 & \cdots & \bar{z}_{N_b} \end{pmatrix}. \tag{9.2}$$

Here, N_b is the number of bands used to discretize the spectral range. We express the observed spectra as an $N_b \times 1$ vector, $\mathbf{b}$. The spectra of the illumination source, $\mathbf{l} = (l_1, l_2, \ldots, l_{N_b})^\top$, is expressed by an $N_b \times N_b$ diagonal matrix, $\mathbf{L}$, such that

$$\mathbf{L} = \mathrm{diag}\,(\mathbf{l}) = \begin{pmatrix} l_1 & 0 & \ldots & 0 \\ 0 & l_2 & \ldots & 0 \\ \vdots & \vdots & \ddots & \vdots \\ 0 & 0 & \ldots & l_{N_b} \end{pmatrix}. \tag{9.3}$$

Here, "diag" represents a function that aligns each element of the vector onto the diagonal elements of a matrix to form the diagonal matrix. We express the spectral reflectance of N_e types of LEDs as an $N_b \times N_e$ matrix, $\mathbf{E}$:

$$\mathbf{E} = \begin{pmatrix} e_{11} & e_{12} & \cdots & e_{1N_e} \\ e_{21} & e_{22} & \cdots & e_{2N_e} \\ \vdots & \vdots & \ddots & \vdots \\ e_{N_b1} & e_{N_b2} & \cdots & e_{N_bN_e} \end{pmatrix}. \tag{9.4}$$

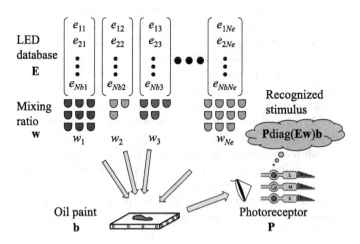

Fig. 9.8 Illumination design using an LED database

We create mixed-light illumination by combining N_e LEDs with N_e mixing ratios. We express the mixing ratios using an $N_e \times 1$ vector, $\mathbf{w}$ (Fig. 9.8). The mixed light can be calculated using a linear summation model:

$$\mathbf{L} = \text{diag}(\mathbf{Ew}). \tag{9.5}$$

Given two paints, paint 1 and paint 2, two types of mixed illumination are referred to as mixed-sources 1 and 2. The method calculates the mixing ratios, such that paints 1 and 2 have the same color and brightness under mixed-source 1, but they appear to have different colors or brightnesses under mixed-source 2. The cost function, $F(\cdot)$, which must be minimized to realize the intended appearance, is as follows:

$$\{\mathbf{w}_1, \mathbf{w}_2\} = \underset{\mathbf{w}_1, \mathbf{w}_2}{\arg\min} \, F(\mathbf{w}_1, \mathbf{w}_2; \mathbf{P}, \mathbf{E}, \mathbf{b}_1, \mathbf{b}_2, \mathbf{u}), \tag{9.6}$$

$$F(\mathbf{w}_1, \mathbf{w}_2; \mathbf{P}, \mathbf{E}, \mathbf{b}_1, \mathbf{b}_2, \mathbf{u}) =$$
$$a_1 \|\mathbf{P}\text{diag}(\mathbf{Ew}_1)\mathbf{b}_1 - \mathbf{P}\text{diag}(\mathbf{Ew}_1)\mathbf{b}_2\|^2$$
$$-a_2 \|\mathbf{P}\text{diag}(\mathbf{Ew}_2)\mathbf{b}_1 - \mathbf{P}\text{diag}(\mathbf{Ew}_2)\mathbf{b}_2\|^{0.5}$$
$$+a_3 \|f(\mathbf{PEw}_1) - f(\mathbf{PEw}_2)\|^2$$
$$+a_4 \max\{\|f(\mathbf{PEw}_1) - \mathbf{u}\|^2, \|f(\mathbf{PEw}_2) - \mathbf{u}\|^2\}, \tag{9.7}$$

where

$$\sum_{n=1}^{N_e} w_{1n} = 1, \quad \sum_{n=1}^{N_e} w_{2n} = 1,$$
$$\sum_{n=1}^{N_e} w_{1n} N_l = N_l, \quad \sum_{n=1}^{N_e} w_{2n} N_l = N_l. \tag{9.8}$$

Moreover, for $n = 1, \ldots, N_e$,

$$w_{1n} \geq 0, \quad w_{2n} \geq 0,$$
$$w_{1n} N_l = \lfloor w_{1n} N_l \rfloor, \quad w_{2n} N_l = \lfloor w_{2n} N_l \rfloor, \tag{9.9}$$

where a_1, a_2, a_3, and a_4 in Eq. (9.7) are nonnegative constants.

We measure the spectra of 53 types ($N_e = 53$) of LED bulbs to form the database, $\mathbf{E}$ (Fig. 9.9). Two colors (i.e., brilliant pink and mars yellow) are used for the oil paints. The method [20] calculated two lights (Figs. 9.10 and 9.11), which satisfies the purpose. Figure 9.12a shows the canvas painted with brilliant pink and mars yellow. Figure 9.12b, c display the canvas illuminated by mixed source 1 and mixed source 2, respectively.

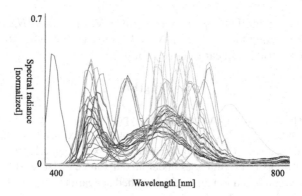

Fig. 9.9 Database of LED spectra. The normalized spectra are shown in this figure, where the spectra having actual brightness were used in our experiments

Fig. 9.10 Appearance of mixed source 1

Fig. 9.11 Appearance of mixed source 2

9.2 Reconstrution of Moving and Deforming Object Shape

The reconstruction of nonrigid objects, (e.g., humans and animals) using active-lighting-based techniques (e.g., 3D scanner) has been intensively researched. Differing from rigid object/scene reconstruction (e.g., [21–23]), which captured static 3D scene from multiple directions and align them, nonrigid objects must be captured in real time, because their shape can change from one frame to the next. One-shot scanning techniques described in Sect. 6.2.2 was used for the solution.

(a) **(b)** **(c)**

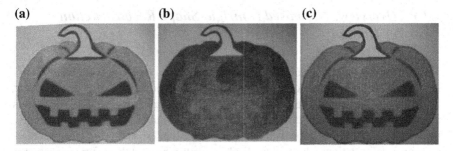

Fig. 9.12 Metamerism art. **a** Two types of oil paint. **b** Appearance under mixed-source 1. **c** Appearance under mixed-source 2

Another challenge for reconstruction of nonrigid objects is shape integration, because rigid object/scene reconstruction can cast 3D reconstruction into rigid-registration problem (e.g., [24–26]), whereas it is difficult for nonrigid objects. For solution, some approaches have addressed this problem without using any prior knowledge of the object [27–30] or by making an assumption about articulated objects (e.g., [31]). These approaches have an inherent weakness when there are unobserved shapes and textures. Another solution is to use prior knowledge of the class of object, if the class is known for captured objects, e.g., a 3D human geometry template is a source of prior knowledge for full-body human shape reconstruction (e.g., [32–35]). Such template-based methods acquire a shape template of the target before actually capturing it in motion, and it subsequently fits the template to measurements obtained from cameras or RGB-depth (RGB-D) sensors [29, 31]. This approach largely relies on nonrigid 3D registration and can suffer from insufficient constraints because of the possible motion of the target objects. Further, human bodies sometimes exhibit large nonrigid deformations according to their poses (e.g., bending arms deforms muscles, skin, and clothes), and thus, statistical template models [36] can represent such pose-dependent deformation only partially.

In our method, full-body reconstruction of moving non-rigid 3D objects, primarily humans, captured by active scanning methods using structured light are explained. In the methods, a statistical model is used for rough reconstruction. Then, loose clothes are handled by estimating pose-dependent deformations in unobserved surfaces, which are represented by a relatively small number of parameters using PCA analysis. Instead of finding an accurate 3D mesh, a rough 3D mesh is precisely deformed, as the base shape of PCA. Thus, pose-dependent deformations (e.g., cloth folds) are efficiently reproduced. Such pose-dependent deformation is called eigen-deformation in the method. To estimate the parameters for eigen-deformation of unobserved body parts, a neural-network (NN)-based coefficient regression that synthesizes deformations for arbitrary poses and viewing directions was proposed [37]. In addition, NN-based estimation of base meshes for eigen-deformation was proposed [38]. Next, details of NN-based methods are explained.

9.2.1   Overview: NN-based Nonrigid Shape Reconstruction of Human

The difficulty in the statistical shape-model-based full-body reconstruction of a moving human body in loose clothes lies mostly in the reproduction of pose-dependent deformations that cannot be described using a statistical shape model. The idea is to represent such deformations via texture and individual displacement of mesh vertices, both of which are embedded into low-dimensional spaces (e.g., eigen-textures and eigen-deformations). Individual displacements represent the difference between the statistical shape-model's mesh and another fully registered to the measurement accounting for relatively large deformations as the texture reproduces the detail. Similar to the the eigen-texture method [39, 40], the system compresses the storage size used for individual textures and displacements by using a small number of eigenvectors and their coefficients. Additionally, it can interpolate unobserved surfaces by using the bases for full body.

Figure 9.13 presents an overview of the full-body reconstruction system. During the preprocessing stage, the system registers a statistic shape model to sequences of point clouds obtained from RGB-D measurements using active scanning methods.

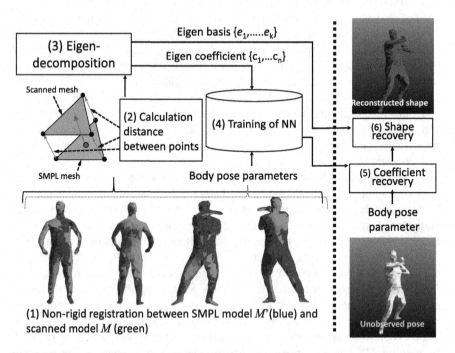

Fig. 9.13 Overview of the system. It first calculates the differences between a scanned shape and a template model. Then, the objects are represented with a small number of eigen-bases and coefficients estimated by human-pose parameters

The skinned multi-person linear (SMPL) model is used for nonrigid registration [35], with which, parameters of the statistical model (i.e., body-shape parameters, $\mathbf{v}$, and a set of joint angles Θ) and the mesh, $\mathbf{M}$, which is fully registered to each point cloud (bottom left of Fig. 9.13), are obtained. As displacements, the difference between $\mathbf{M}'$ and $\mathbf{M}$, which can be measured solely from Θ, is computed. Then, the displacements are embedded into a low-dimensional subspace via eigen-deformation using a similar method as the eigen-texture method (top-left of Fig. 9.13). During the rendering stage, $\mathbf{M}'$ is first recovered from Θ, and the textures and displacements are then reconstructed from the coefficients, which are estimated by using an NN-based regression. Adding all displacements to $\mathbf{M}'$, the clothing mesh, $\mathbf{M}$, is reconstructed.

9.2.2 Rough Representation of Human Body Shape Using SMPL Model

For rough shape representation of the human body, the system uses the SMPL model [35] to the input sequence of point clouds, each of which is a human body scan. This facilitates compression and efficient representation of rough human body shapes wearing loose clothing, because the SMPL model gives us an efficient parametrization of the human body using the small number of body-shape parameters β and joint angles θ by minimizing the following energy function:

$$E(\beta, \theta) = \sum_{i \in C_{\text{anc}}} \|x_i'(\beta, \theta) - y_i\|^2$$
$$+ P(\theta) + R(\theta) + S(\beta) + Q(\beta, \theta), \qquad (9.10)$$

where $x_i'(\beta, \theta)$ and y_i are the ith vertex in $M'(\beta, \Theta)$ and its corresponding point in $\mathbf{T}$. P is a pose prior built as the logarithm of as the sum of a mixture of Gaussians, which keeps θ in probable pose space. R is the penalties for unusual joint angles, S represent large β values that correspond to implausible shapes, and Q represents inter-penetration of the body parts. More details on these penalty terms can be found in [41].

Second, the skin-tight 3D mesh, $\mathbf{M}'$, is inflated to roughly minimize the distance between the mesh of $\mathbf{M}'$ and the target cloud points, $\mathbf{T}$, according to their silhouette points. The silhouette points are identified in 2D reprojected images. When using a single camera, we project both $\mathbf{M}'$ and $\mathbf{T}$ to the image plane of the camera to find the silhouette points. This step introduces more flexibility to the vertex positions so that $\mathbf{M}$ can be closer to $\mathbf{T}$, making the final fitting step more stable.

In this step, we restrict the possible displacement of the ith vertex in $\mathbf{M}$ to

$$x_i(\beta, \theta, \{n, l\}) = x_i'(\beta, \theta) + n_i l_i, \qquad (9.11)$$

for all i, where $x_i \in M$, $x'_i \in M'$, n_i is the direction of the movement on M' at x', and l_i is the amplitude of the displacement. Here, Kimura et al. [37] allowed each vertex to move only in their normal direction, whereas we allow for more freedom in the displacement direction. This new formulation prevents each mesh to interpenetrate (e.g., around the crotch when a person stands). The energy function to be minimized according to the silhouette correspondences, C_{sil}, during the fitting process is

$$
E(\{n, l\}) = \sum_{i,j \in C_{sil}} \|x_i - y_j\|^2 + \lambda_l \sum_{i,j \in A_{ver}} (l_i - l_j)^2
$$
$$
+ \lambda_{normal} \sum_{i,j \in A_{face}} |N_i^{face} - N_j^{face}|
$$
$$
+ \lambda_{rot} \sum |n_i - n'_i| + \lambda_{laplacian} \sum |x_i - \frac{1}{z_i} \sum_{j \in \mathcal{N}_i} x_j|, \quad (9.12)
$$

where A_{ver} and A_{face} are adjacency of vertices and faces of the SMPL model, respectively. N_i^{face} is the normal of the ith face. n'_i is the normal vectors of the ith vertex of M'. $\mathcal{N}_i$ is the set of adjacent indices of the ith vertex, and z_i is the number of adjacent vertices around the ith vertex. λ_l, λ_{normal}, λ_{rot}, and $\lambda_{laplacian}$ are weights used to control the contributions of the associated terms. The second term is a regularizer to keep the displacement of the adjacent vertices similar. The third term regularizes the direction of the adjacent mesh normals to avoid overly flat surfaces. The fourth term keeps the vertex direction of the movement as rigid as possible. The last term is the Laplacian mesh regularizer inspired from [42, 43], which enforces smoothness.

At the final step, all vertices and points instead of the silhouettes are used to make correspondences. Additionally, the displacements are not restricted by Eq. (9.11). Thus,

$$
x_i(\boldsymbol{\beta}, \boldsymbol{\theta}, \{d\}) = x'_i(\boldsymbol{\beta}, \boldsymbol{\theta}) + d_i. \quad (9.13)
$$

With these modifications, the energy function to be minimized is

$$
E(\{d\}) = \sum_{i \in C_{all}} \|x_i - y_i\|^2 + \lambda_{normal} \sum_{i,j \in A_{face}} |N_i^{face} - N_j^{face}|
$$
$$
+ \lambda_{laplacian} \sum |x_i - \frac{1}{z_i} \sum_{j \in \mathcal{N}_i} x_j|, \quad (9.14)
$$

where C_{all} are the correspondences between M and T. In this step, the nearest neighbors from T to M are updated iteratively. At the final iteration, if the target point cloud T is a full-body human datum, we identify correspondences, not only from T to M, but also from M to T. This allows us to fill gaps, which sometimes occur around the armpit and the crotch. Owing to the Laplacian regularizer term, vertices in the SMPL model that do not have any correspondence also move by being dragged by the other vertices. After this nonrigid registration step, the body-shape parameters, β, the joint angles, θ, and a fully registered 3D mesh, $\mathbf{M}$, for each point cloud in the input sequence can be obtained.

9.2.3 Detail Representation of Human Body Shape Using PCA

To synthesize details of shape deformations and unobserved surfaces and to compress the storage of individual vertex positions, eigen-deformation is proposed. The basic idea is almost the same as eigen-texture [39, 40]. However, eigen-deformation deals with vertex positions. There is an inherent difference between textures and deformations. Given that a mesh is well registered to the point cloud, the variations in textures are not significant, because such variations are caused solely by local deformations, such as wrinkles. On the other hand, those in vertices come from body poses. Thus, changes in the shoulder-joint angle result in large changes in the vertex positions of forearms. Therefore, the direct application of eigen-decomposition to vertex positions may not work well.

To improve the representativity, displacement vectors of each part between the statistic model mesh, $\mathbf{M}'$, and the fully registered mesh, $\mathbf{M}$, are computed. This represents the displacement vector in a certain coordinate system associated with each body part. Therefore, only the difference between $\mathbf{M}'$ and $\mathbf{M}$ is counted in the displacement vector.

The displacement vector, $\mathbf{q_k}$, of the kth vertex of the lth body part in mesh $\mathbf{M}$ are computed by $q_{lk} = H_l(v_{lk} - v'_{lk})$ for the corresponding vertex position in $\mathbf{M}'$, where v_{lk} and v'_{lk} are the kth vertices in the lth body parts in $\mathbf{M}$ and $\mathbf{M}'$, and H_l is a rigid transformation matrix between the entire body coordinate system and each body part's system. These displacement vectors are concatenated to form a column vector, $\mathbf{q}_l^\top = (q_{l1}\, q_{l2} \ldots)^\top$. Then, these displacement vectors are aggregated over all frames in which the triangle is visible. These vectors are centralized as with the eigen-texture and again concatenated to form matrix $\bar{Q}_l$. Then, eigen-decomposition to $\bar{Q}_l\bar{Q}_l^\top$ is conducted to obtain the eigenvectors. The displacement vectors are further embedded/reconstructed into/from the subspace spanned by the eigenvectors. When reconstructing $\mathbf{M}$, we first recover $\mathbf{M}'$ from Θ and v using Eq. (9.14), and we recover $\tilde{\mathbf{q}}_k$ using a small number of eigenvectors. Finally, $H_l^{-1}\tilde{\mathbf{q}}_k$ is calculated for each vertex of each body part and added to $\mathbf{M}'$.

9.2.4 Convolutional Recurrent NNs (CRNN) for Coefficient Regression

To enable full-body reconstruction, unobserved deformations in unobserved surfaces should be interpolated. This is accomplished using coefficient regression in the eigen-deformation spaces. Provided that RGB-D sensors are fixed, the variations of the displacements of the same triangle are solely explained by the person's pose. More specifically, joint angles represented by rotation matrices mostly determine them. This implies that coefficients for eigenvectors or the coordinates in the low-dimensional spaces can be regressed from the rotation matrices. Therefore, it can be trained by NN-based regressors that map a joint angle (i.e., rotation matrix) to the coefficients.

Let $\mathbf{r} \in \mathbb{R}^9$ be the vectorization of the rotation matrix that represents a certain body part's joint angle. Because the relationship between rotation matrices and coefficients are unknown, we use an NN having two layers to represent the nonlinearity. Our regressor gives coefficients $\tilde{c} \in \mathbb{R}^L$ by

$$\tilde{c}(r) = W_2 \tanh(W_1\mathbf{r} + b_1) + b_2, \tag{9.15}$$

where W_1 is in $\mathbb{R}^{J \times 9}$ and W_2 is in $\mathbb{R}^{L \times J}$. The regressor is trained using the gradient-descent algorithm. For regularization, we employ weight decay. Examples of the estimated coefficient from poses are shown in Fig. 9.14, and reconstructed shapes are shown in Fig. 9.15, implying authentication of the algorithm.

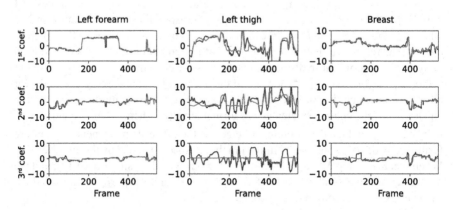

Fig. 9.14 Coefficient of the eigenvalue of original and estimated data

Fig. 9.15 RMSE of eigen-deformations using 10 regressed coefficients. Although some parts have large errors, they are from nonrigid registration failures. Thus, few artifacts are observed in final rendering results with textures shown in Fig. 9.28

0.03 m

0 m

9.2.5 Binary-Ensemble NN (BENN) for Base-Mesh Estimation

PCA-based base-mesh estimation can be applied only if the entire human body shape is captured. To overcome this limitation, NN-based base-mesh estimation methods have been proposed. Figure 9.16 shows an overview of such a method. During the regression stage, a human pose and regress displacements, D, are used as inputs. BENN is a network with bases of the displacements as part of weight parameters, and it directly regresses displacements, $D(\{\Theta\})$. When the training data is partially occluded, a visibility map is used to ignore the occluded parts.

In actual implementation, BENN is composed of the CRNN with an additional output layer. In the layer, the bases and the means calculated in the previous section are used as weight and bias parameters are updated in training. This allows the network

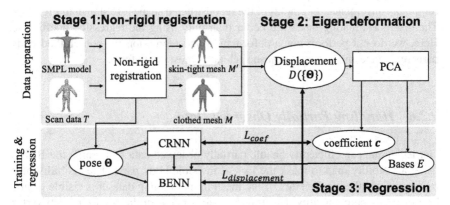

Fig. 9.16 Overview of our system. We first calculate the displacements between human shape with clothes and a template model. Then, these displacements are represented by a small number of bases and coefficients. The displacements and coefficients are estimated by regressing the human pose parameters

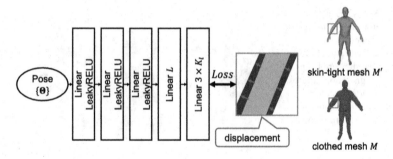

Fig. 9.17 Network architecture of BENN. When the target data are partially observed, we use visibility map not to calculate the gradient of the unobserved body part

to output more accurate displacements to directly regress the displacements. The bases, E, and the mean shape, $\overline{d}$, obtained at the end of the eigen-deformation stage are used as the initial weights and biases of the additional output layer. The inputs of the BENN are the same as those of the CRNN. The output of the BENN are the displacements, $\tilde{d} \in \mathbb{R}^{(3 \times K_l)}$, of all vertices that belonging to lth body part, where K_l is the number of vertices. The bases are determined by the output of the BENN and the means of all bases are equal to the bias of the BENN (Fig. 9.17). Finally, the displacement values are computed as

$$\tilde{d}(\Theta) = W\tilde{c}(\Theta) + \mathbf{b}, \tag{9.16}$$

where W is in $\mathbb{R}^{(3 \times K_l) \times L}$ and $\mathbf{b}$ is in $\mathbb{R}^{(3 \times K_l)}$.

Next, when the BENN is trained with partially observed data, the previous step will be omitted. However, the previous step is still necessary to obtain a good initial value for quick convergence. Examples of the reconstructed shapes are shown in Fig. 9.18. The results estimating bases are better than only using the regression coefficient. Here, we do not perform eigen-deformation and regression for non-clothed body parts (e.g., face and hands), because such parts must not deform.

9.2.6 Handling Partially Observed Data

Our technique can efficiently handle partially observed data by training the BENN using a visibility map to mask the loss of invisible body parts. The visibility map represents whether each vertex of the mesh in the training dataset is visible or not. The map can be created by detecting collisions between the ray from each vertex to the principal point and each mesh. This allows us to back propagate only from the unit of the output layer corresponding to the visible body vertex. To utilize this mask-based method, subjects have to show all body parts at least once.

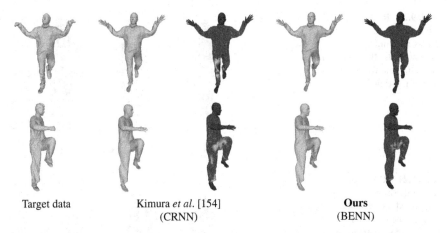

Target data   Kimura *et al.* [154]   **Ours**
                            (CRNN)   (BENN)

Fig. 9.18 Qualitative comparison using public dataset. The left-most column shows target data, the next two columns show the results of Kimura et al. [37] (CRNN), and the last two column show the results of our proposed method (BENN). The second column of each result shows the error in color. (Blue = 0 mm and red >= 50 mm) (Color figure in online)

9.2.7 Experiments

Evaluation of CRNN Using Synthetic Data

CRNN was evaluated using synthetic data with a commercially available 3D mesh model of the entire human body with and without clothes. A skeleton (bones) was attached to the mesh model so that sequences of 3D meshes could be created by using 3D-CG software (e.g., 3D-MAX). Because muscle deformation and cloth simulation were employed during rendering, realistic shape deformations having complicated shading effects were represented. Some examples of rendered images are shown in Fig. 9.19.

First, cumulative contribution ratios are shown in Fig. 9.22, and differences from ground-truth meshes (i.e., clothed meshes) are shown in Fig. 9.20. We can see that

Fig. 9.19 Synthetic data for evaluation

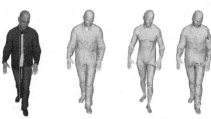

Textured   Non-textured   Naked   Clothed
model   model   model   model
(sim data) (sim data)   (SMPL) (SMPL)

Fig. 9.20 RMSE of eigen-deformation using a different number of components. 10 components are sufficient for error-free reconstruction

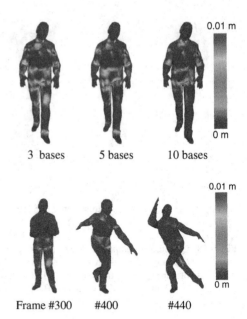

3 bases   5 bases   10 bases

Fig. 9.21 RMSE of eigen-deformation using 10 regressed coefficients. Unlike the real data (Fig. 9.15), all errors are small, because simulation data has little failure on non-rigid registration

Frame #300   #400   #440

10 eigenvectors were sufficient to represent the original mesh. This is equivalent to 2.52% of the original data.

Next, regression results are shown in Fig. 9.21. NNs worked well with synthetic data and most of the body parts had small errors.

Finally, two scenarios having short- and long-term interpolations (Fig. 9.23) were conducted to evaluate interpolation accuracy. Results of interpolation and extrapolation are shown in Figs. 9.24 and 9.25 and coefficient values of extrapolation are shown in Fig. 9.26. Extrapolation tended to create larger errors than interpolation. However, regressed coefficients still had a similar trend to ground truth. The textured

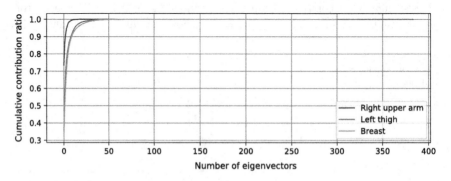

Fig. 9.22 Cumulative contribution ratios of three example body parts showing only small numbers of eigen-bases contributing to the original deformation

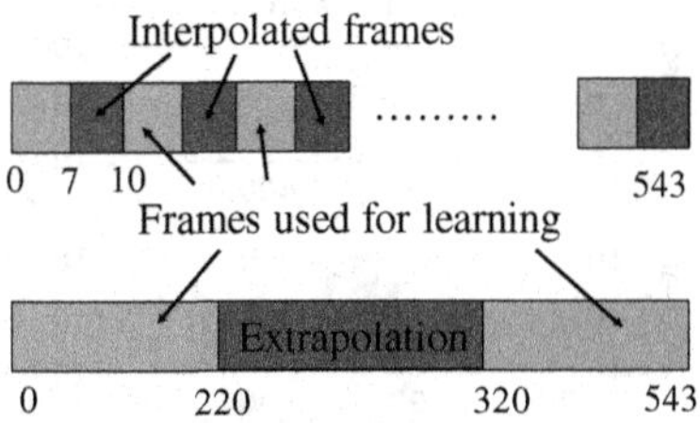

Fig. 9.23 Frames interpolated and extrapolated for experiments

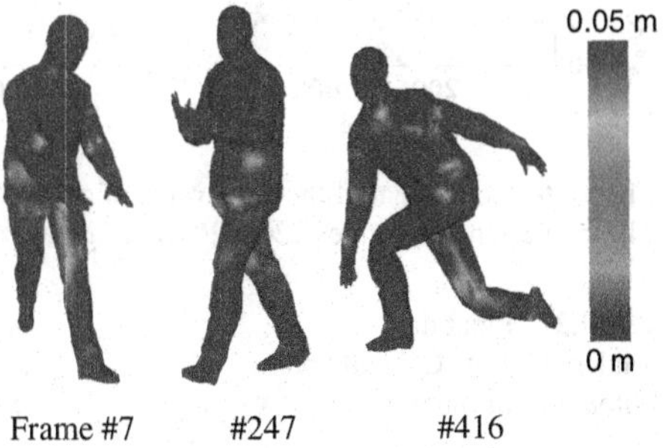

Fig. 9.24 Interpolation results by recovering 10 components by regression. RMSE errors are all as small as original ones

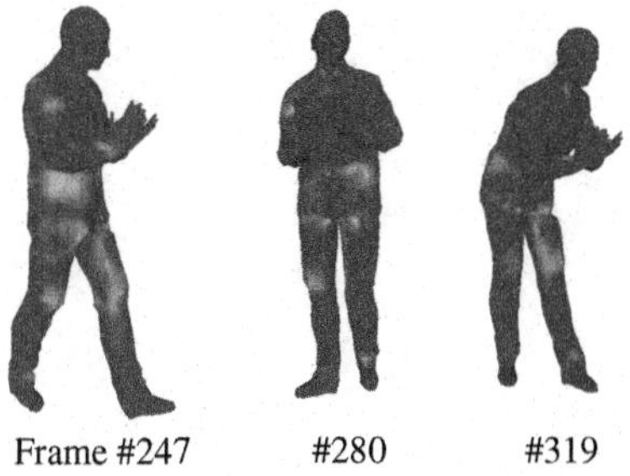

Frame #7   #247   #416

Fig. 9.25 Extrapolation results by recovering 10 components by regression. Even almost new pose is reconstructed, RMSE errors are still small

Frame #247   #280   #319

results with interpolated shapes are shown in Fig. 9.27. Considering the compression rate of 2.52%, the visual quality of the rendered images is comparable to video compression in addition to the significant advantage of arbitrary viewpoint rendering.

Demonstration of CRNN with Real Data

Two calibrated RGB-D sensors were used to obtain two sequences of a moving person from the person's front and back in a real data experiment. A pair of depth measurements from each corresponding pair of frames were integrated according to the RGB-D sensors' relative poses to create a single point cloud. Although this point cloud and the corresponding RGB images had unobserved (but not large) surfaces because of self-occlusion, the proposed technique synthesized the deformations of such surfaces.

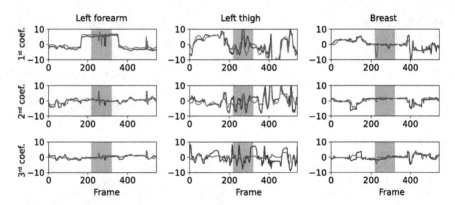

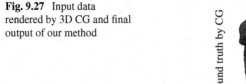

Fig. 9.26 Coefficient of the eigenvalue of original data and the estimated data with extrapolation. The coefficients in frames 220–320 (highlighted) are excluded during training

Fig. 9.27 Input data rendered by 3D CG and final output of our method

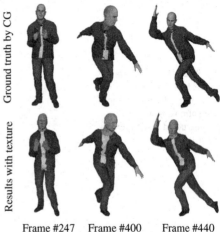

Frame #247   Frame #400   Frame #440

Regression estimation results are shown in Fig. 9.15. These results demonstrate that 10 eigenvectors were sufficient as well for the real data case. Final rendering results are shown in Fig. 9.28, implying the effectiveness of the approach.

Evaluation of Inter-frame Interpolation by CRNN and BENN

The reconstruction results obtained from CRNN, BENN, and the previous method [44] using the publicly available dataset [45], consisting of human mesh with loose clothes, are compared. In our method, we fitted the model to the target scan data and regressed the clothes deformations. On the other hand, Yang et al. [44] calculated the displacements between models and target mesh without a fitting strategy. Instead, they assumed that the input scan data had mesh consistency.

Following Yang et al. [44], 80% of each sequence was used for training and others for testing. Although Yang et al. used 40 bases for regression, CRNN and BENN

captured images by Kinect   Kinect   Proposed

Frame #0

Kinect   Proposed   Kinect   Proposed

Frame #200   Frame #400

Fig. 9.28 Real input data (i.e., captured image and 3D mesh by Kinect) and the final results of our method. Note that our method only uses a tight-clothed template model, but successfully recovers loose-wear (pants)

require only 30 bases. The quantitative comparison result is shown in Table 9.1. In this table, we show the results of BENN and the average RMSE in *mm* of each vertex. Because the density of vertices differs between the target point clouds and our outputs, we could not simply calculate the error using one-to-one correspondences of points in-between. Therefore, the error between clothed meshes obtained at the nonrigid registration stage (Sect. 9.2.2) and the results of the proposed method were calculated. BENN outperformed previous methods in most sequences while using fewer bases than Yang et al. and the same amount as Kimura et al. (CRNN). A qualitative comparison is shown in Fig. 9.18. The first column shows the target data of nonrigid registration. The next two columns show the results obtained with the method of Kimura et al. (CRNN) and the right two columns show the results obtained with our proposed method. The first column of each set of two columns shows the results of regressed deformation and the second shows the error with pseudo-color (blue = 0 mm and red >= 50 mm). From these results, it is confirmed that BENN also worked well qualitatively.

Table 9.1 Average RMSE of each vertex position in mm. Ours reflect the result of BENN. (The value of first row is cited from [44])

Seq	Bounc.	Hand.	Crane	Jump.	Mar. 1	Mar. 2	Squ. 1	Squ. 2
Yang et al. [44]	10.27	–	**4.27**	–	3.93	–	4.31	–
Kimura et al. [37] (CRNN)	5.45	3.45	7.61	**9.24**	3.06	8.25	3.20	7.36
(BENN)	**5.28**	**3.03**	7.70	9.52	**2.87**	**8.13**	**2.87**	**7.29**

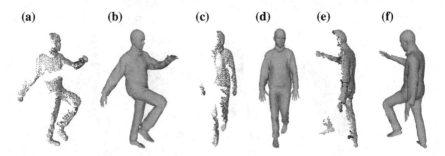

Fig. 9.29 **a, c, d**: the target scan data of nonrigid registration. **b, d, f**: the results of intra-frame interpolation. Even at the invisible body part, the clothes deformations are reconstructed well

Demonstration of Intra-frame Interpolation by BENN

Owing to BENN implementation, invisible parts of the scanned data were efficiently handled. This allows us to use partially observed data to train the network. Intra-frame interpolation of the partially observed data generated from the public dataset [45] was conducted. Furthermore, a sequence of RGB-D images of a human in motion using Kinect-v2 was captured and intra-frame interpolation was performed to compare the results obtained with the proposed method to those obtained with DoubleFusion [46], which is a real-time system that reconstructed not only the human inner body via optimized parameters of the SMPL model. It also reconstructed the clothes worn by the subject with a single-depth camera (Kinect-v2). The results are shown in Figs. 9.29 and 9.30. In Fig. 9.30, the first column shows the target data of nonrigid registration of our proposed method and the input data for DoubleFusion. Invisible body parts were reconstructed during the regression step, revealing the effectiveness of the proposed method to handle partially observed data.

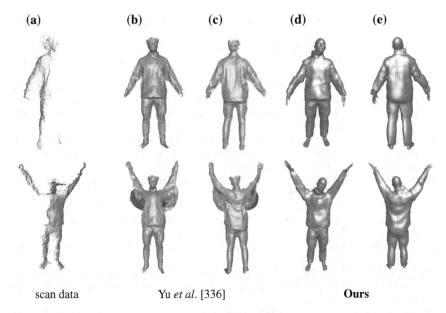

(a) **(b)** **(c)** **(d)** **(e)**

scan data   Yu *et al.* [336]   **Ours**

Fig. 9.30 Qualitative comparison with [46] using Kinect-v2. **b** and **d**: results of two methods showing the front view. **c** and **e**: results of the backside view. DoubleFusion resulted in some artifacts (e.g., armpit), whereas there were no artifacts in ours, since our method can represent pose-dependent deformation

References

1. Raskar R, Welch G, Low KL, Bandyopadhyay D (2001) Shader lamps: animating real objects with image-based illumination. In: Rendering techniques. Springer, pp 89–102
2. Mine MR, Van Baar J, Grundhofer A, Rose D, Yang B (2012) Projection-based augmented reality in disney theme parks. Computer 45(7):32–40
3. Bimber O, Gatesy SM, Witmer LM, Raskar R, Encarnação LM (2002) Merging fossil specimens with computer-generated information. Computer 9:25–30
4. Mukaigawa Y, Nishiyama M, Shakunaga T (2004) Virtual photometric environment using projector. In: Proceedings of the international conference on virtual systems and multimedia, pp 544–553
5. Hamasaki T, Itoh Y, Hiroi Y, Iwai D, Sugimoto M (2018) Hysar: hybrid material rendering by an optical see-through head-mounted display with spatial augmented reality projection. IEEE Trans Vis Comput Graph 24(4):1457–1466
6. Raskar R, Beardsley P, van Baar J, Wang Y, Dietz P, Lee J, Leigh D, Willwacher T (2004) Rfig lamps: interacting with a self-describing world via photosensing wireless tags and projectors. In: ACM transactions on graphics (TOG), vol 23. ACM, pp 406–415
7. Hiura S, Tojo K, Inokuchi S (2003) 3-d tele-direction interface using video projector. In: ACM SIGGRAPH 2003 sketches & applications. ACM, pp 1–1
8. Kawabe T, Fukiage T, Sawayama M, Nishida S (2016) Deformation lamps: a projection technique to make static objects perceptually dynamic. ACM Trans Appl Percept (TAP) 13(2):10
9. Punpongsanon P, Iwai D, Sato K (2018) Flexeen: visually manipulating perceived fabric bending stiffness in spatial augmented reality. IEEE Trans Vis Comput Graph

10. Iwai D, Aoki M, Sato K (2018) Non-contact thermo-visual augmentation by IR-RGB projection. IEEE Trans Vis Comput Graph 25(4):1707–1716
11. Punpongsanon P, Iwai D, Sato K (2015) Projection-based visualization of tangential deformation of nonrigid surface by deformation estimation using infrared texture. Virtual Real 19(1):45–56
12. Punpongsanon P, Iwai D, Sato K (2015) Softar: visually manipulating haptic softness perception in spatial augmented reality. IEEE Trans Vis Comput Graph 21(11):1279–1288
13. Kanamori T, Iwai D, Sato K (2018) Pseudo-shape sensation by stereoscopic projection mapping. IEEE Access 6:40649–40655
14. Mine R, Iwai D, Hiura S, Sato K (2017) Shape optimization of fabricated transparent layer for pixel density uniformalization in non-planar rear projection. In: Proceedings of the 1st annual ACM symposium on computational fabrication. ACM, p 16
15. Asayama H, Iwai D, Sato K (2017) Fabricating diminishable visual markers for geometric registration in projection mapping. IEEE Trans Vis Comput Graph 24(2):1091–1102
16. Amano T (2012) Shading illusion: a novel way for 3-d representation on paper media. In: Proceedings of the IEEE conference on computer vision and pattern recognition (CVPR) workshops. IEEE, pp 1–6
17. Okutani N, Takezawa T, Iwai D, Sato K (2018) Stereoscopic capture in projection mapping. IEEE Access 6:65894–65900
18. Park JI, Lee MH, Grossberg MD, Nayar SK (2007) Multispectral imaging using multiplexed illumination. In: Proceedings of the international conference on computer vision (ICCV). IEEE, pp 1–8
19. Bala R, Braun KM, Loce RP (2009) Watermark encoding and detection using narrowband illumination. In: Color and imaging conference, vol 2009. Society for Imaging Science and Technology, pp 139–142
20. Miyazaki D, Nakamura M, Baba M, Furukawa R, Hiura S (2016) Optimization of led illumination for generating metamerism. J Imaging Sci Technol 60(6):1–60502
21. Jancosek M, Pajdla T (2011) Multi-view reconstruction preserving weakly-supported surfaces. In: Proceedings of the IEEE conference on computer vision and pattern recognition (CVPR), pp 3121–3128
22. Furukawa Y, Ponce J (2009) Accurate, dense, and robust multiview stereopsis. IEEE Trans Pattern Anal Mach Intell (PAMI) 32(8):1362–1376
23. Besl PJ, McKay ND (1992) Method for registration of 3-d shapes. In: Sensor fusion IV: control paradigms and data structures, vol 1611. International Society for Optics and Photonics, pp 586–606
24. Izadi S, Kim D, Hilliges O, Molyneaux D, Newcombe R, Kohli P, Shotton J, Hodges S, Freeman D, Davison A, Fitzgibbon A (2011) Kinectfusion: real-time 3d reconstruction and interaction using a moving depth camera. In: Proceedings of the ACM symposium on user interface software and technology (UIST), pp 559–568
25. Newcombe RA, Izadi S, Hilliges O, Molyneaux D, Kim D, Davison AJ, Kohi P, Shotton J, Hodges S, Fitzgibbon A (2011) Kinectfusion: real-time dense surface mapping and tracking. In: 2011 10th IEEE international symposium on mixed and augmented reality. IEEE, pp 127–136
26. Whelan T, McDonald J, Kaess M, Fallon M, Johannsson H, Leonard J (2010) Kintinuous: Spatially extended KinectFusion. In: Proceedings of the RSS workshop on RGB-D: advanced reasoning with depth cameras
27. Amberg B, Romdhani S, Vetter T (2007) Optimal step nonrigid ICP algorithms for surface registration. In: Proceedings of the IEEE conference on computer vision and pattern recognition (CVPR), pp 1–8
28. Li H, Sumner RW, Pauly M (2008) Global correspondence optimization for non-rigid registration of depth scans, pp 1421–1430
29. Zollhöfer M, Nießner M, Izadi S, Rehmann C, Zach C, Fisher M, Wu C, Fitzgibbon A, Loop C, Theobalt C et al (2014) Real-time non-rigid reconstruction using an RGB-D camera. ACM Trans Graph (TOG) 33(4):1–12

30. Newcombe RA, Fox D, Seitz SM (2015) Dynamicfusion: reconstruction and tracking of non-rigid scenes in real-time. In: Proceedings of the IEEE conference on computer vision and pattern recognition (CVPR), pp 343–352

31. Li H, Adams B, Guibas LJ, Pauly, M.: Robust single-view geometry and motion reconstruction. ACM Trans Graph (TOG) 28(5):175:1–175:10

32. Anguelov D, Srinivasan P, Koller D, Thrun S, Rodgers J, Davis J (2005) Scape: shape completion and animation of people. In: ACM transactions on graphics (TOG), vol 24. ACM, pp 408–416

33. Chen Y, Liu Z, Zhang Z (2013) Tensor-based human body modeling. In: Proceedings of the IEEE conference on computer vision and pattern recognition (CVPR), pp 105–112

34. Bogo F, Black MJ, Loper M, Romero J (2015) Detailed full-body reconstructions of moving people from monocular RGB-D sequences. In: Proceedings of the international conference on computer vision (ICCV), pp 2300–2308

35. Loper M, Mahmood N, Romero J, Pons-Moll G, Black MJ (2015) Smpl: a skinned multi-person linear model. ACM Trans Graph (TOG) 34(6):1–16

36. Pishchulin L, Wuhrer S, Helten T, Theobalt C, Schiele B (2015) Building statistical shape spaces for 3d human modeling, pp 1–10. arXiv:1503.05860

37. Kimura R, Sayo A, Dayrit FL, Nakashima Y, Kawasaki H, Blanco A, Ikeuchi K (2018) Representing a partially observed non-rigid 3d human using eigen-texture and eigen-deformation. In: Proceedings of the international conference on pattern recognition (ICPR). IEEE, pp 1043–1048

38. Sayo A, Onizuka H, Thomas D, Nakashima Y, Kawasaki H, Ikeuchi K (2019) Human shape reconstruction with loose clothes from partially observed data by pose specific deformation. In: Pacific-Rim symposium on image and video technology (PSVIT). Springer, pp 225–239

39. Nishino K, Sato Y, Ikeuchi K (2002) Eigen-texture method: appearance compression and synthesis based on a 3d model. IEEE Trans Pattern Anal Mach Intell (PAMI) 23(11):1257–1265

40. Nakashima Y, Okura F, Kawai N, Kimura R, Kawasaki H, Ikeuchi K, Blanco A (2017) Realtime novel view synthesis with eigen-texture regression. In: Proceedings of the British machine vision conference (BMVC)

41. Bogo F, Kanazawa A, Lassner C, Gehler P, Romero J, Black MJ (2016) Keep it smpl: automatic estimation of 3d human pose and shape from a single image. In: Proceedings of the European conference on computer vision (ECCV). Springer, pp 561–578

42. Sorkine O, Cohen-Or D, Lipman Y, Alexa M, Rössl C, Seidel HP (2004) Laplacian surface editing. In: Proceedings of the 2004 Eurographics/ACM SIGGRAPH symposium on Geometry processing, pp 175–184

43. Nealen A, Igarashi T, Sorkine O, Alexa M (2006) Laplacian mesh optimization. In: Proceedings of the 4th international conference on computer graphics and interactive techniques in Australasia and Southeast Asia, pp 381–389

44. Yang J, Franco JS, Hétroy-Wheeler F, Wuhrer S (2018) Analyzing clothing layer deformation statistics of 3d human motions. In: Proceedings of the European conference on computer vision (ECCV), pp 237–253

45. Vlasic D, Baran I, Matusik W, Popović J (2008) Articulated mesh animation from multi-view silhouettes. ACM Trans Graph (TOG) 27(3):97:1–97:9. https://doi.org/10.1145/1360612.1360696

46. Yu T, Zheng Z, Guo K, Zhao J, Dai Q, Li H, Pons-Moll G, Liu Y (2018) Doublefusion: real-time capture of human performances with inner body shapes from a single depth sensor. In: Proceedings of the IEEE conference on computer vision and pattern recognition (CVPR), pp 7287–7296

Chapter 10
Biomedical Application

For medical diagnoses and treatments, basic and clear information, including heart rate, blood pressure, X-ray imagery, etc., of a subject is essential. Thus, a huge amount of sensing technologies have been researched, developed, and commercialized. Among the wide variety of available sensors (e.g., electromagnetic (i.e., MRI), radiation- based (i.e., X-ray), and ultrasound), active-lighting-based techniques have recently drawn a great deal of attention because of their noninvasive methods and simple configurations. In this chapter, a 3D endoscope system using structured light and a noncontact heartbeat detection system is described.

10.1 3D Endoscope

Endoscopic diagnoses and treatments of digestive-tract tumors have become popular, because such procedures facilitate early-stage diagnoses and minimally invasive surgeries. Currently, accurate tumor-size measurement is an open problem for the effectiveness of practical endoscopic systems. To improve measurement accuracy, 3D endoscopic systems have been investigated. These systems employ various techniques such as SFS [1–3], binocular stereo vision [4], photometric stereo [5], single-line laser scanner attached to the head of a scope [6], and one-shot structured light based systems [7–11].

Constructing a 3D endoscope based on active-lighting techniques is a promising solution because of its simple configuration and potentially high accuracy. By simply adding a pattern projector to a common monocular endoscope, we can capture 3D information inside the human body. Of course, there are challenging issues in this approach that must be overcome. The qualities of images captured by endoscopes are much lower than those of industrial cameras because of size limitations, such as small lenses. Thus, the captured images are normally affected by specularities and

© Springer Nature Switzerland AG 2020
K. Ikeuchi et al., *Active Lighting and Its Application for Computer Vision*,
Advances in Computer Vision and Pattern Recognition,
https://doi.org/10.1007/978-3-030-56577-0_10

strong subsurface scatterings, which are common in bio-tissues. To date, to overcome the problems, several solutions have been recently proposed. Next, a typical 3D endoscopic system based on an active-lighting technique, is explained.

10.1.1   Ultra-Small One-Shot Structured Light System for the Endoscope

We developed an endoscopic system that enables 3D measurement [7], based on active-stereo techniques using the projection of a static pattern [12]. We use a normal monocular endoscope, in which there was an instrument channel, wherein endoscopists could insert instruments for medical treatments. We built a micro-sized pattern projector that could be inserted into this channel so that a projector-camera system could be formed.

Figure 10.1 shows the prototype configuration of the proposed system. The endoscope system used for our system comprised a FUJIFILM VP-4450HD and an EG-590WR. As shown in the figure, the pattern projector was shaped as a small cylinder having a diameter of about 2.8 mm mounted at the tip of an optical fiber. The light source of the pattern projector was a green laser module having a wavelength of 532 nm. The laser light from the light source was transferred via a plastic optical fiber (POF) and led to the head of the pattern projector. The device could be either inserted through the instrument channel (Fig. 10.10a) or tightly attached to the head of an endoscope (Fig. 10.1).

To date, we have developed two versions of the micro-projector. The first version was based on the projection of a pattern-printed film, including slide projectors as shown in Fig. 10.1. In this version, a micro-pattern chip, on which the pattern image was printed, was set at the tip of the POF. Lights passing through the pattern chip also passed through the nonspherical lens placed in front of the chip and was emitted to the target surface. According to the optical effect, the pattern on the chip was projected onto the target. As shown in Fig. 10.2a, the pattern consisted of a grid pattern having vertical and horizontal wave lines. Figure 10.2b shows the actual image, whereon the pattern was projected onto a planar object. By setting the wavelength of the horizontal wave pattern to not be an integral multiplication of the horizontal interval of the vertical wave pattern, the pattern becomes locally unique for each grid point within the length of the liquid-crystal matrix (LCM). Thus, the local appearance was used to find correspondences between the camera image and the projected pattern.

The first version of the projector had some problems. One problem was related to the shallow depth of field. It is well known that the depths of fields of video or slide projectors are generally narrower than those of cameras. Similarly, the in-focus range of the first-version projector was approximately 8 mm from a 40-mm working distance, mainly because it used a single nonspherical thin lens to project a focused image of the film-printed pattern. Another problem was that the system brightness

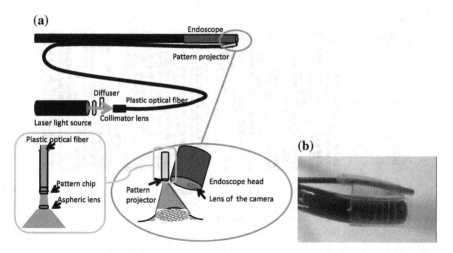

Fig. 10.1 3D endoscope system with microstructure light laser project. **a** Schematic diagram. **b** Actual endoscope head attached with laser projector fixed by Impact-ShooterTM^{TM}

Fig. 10.2 Endoscope pattern. **a** Pattern image (wave-grid pattern). **b** A projected grid pattern

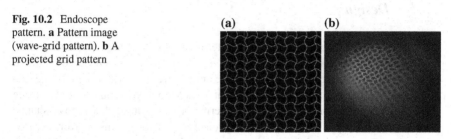

was insufficient, especially given the reflectance properties of green wavelengths on internal tissue. This resulted in poor reconstruction quality.

To solve both problems, we developed a second-version projector that used DOE. Apart from the DOE, it comprised a grin lens, a single-mode optical fiber, and a laser light source, as shown in Fig. 10.3a. Because the DOE can project a fine, complex pattern at a greater depth range without requiring lenses, the energy loss was less than 5%. The actual specifications of the micro-pattern projector are as follows. To lead the micro DOE projector into the head of the endoscope through the instrument channel, its dimensions were kept to be 2.8 mm in diameter and 12 mm in length. The working distance was about 20–30 mm, and the area of the projected pattern was about 20 × 20 mm for a 20-mm working distance.

For the second-version projector, a novel pattern designed for endoscope environments was adopted. The design approach is described in the following subsection.

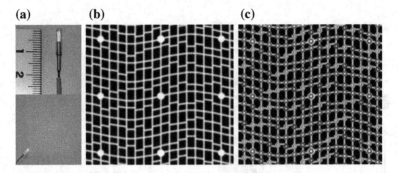

(a)   **(b)**   **(c)**

Fig. 10.3 DOE micro-projector and its pattern with embedded codewords. **a** DOE projector. **b** Projected pattern. **c** Embedded codewords of S colored in red, L in blue, and R in green. S indicates that the edges of the left and the right sides have the same height, L means the left side is higher, and R means the right is higher (Color figure online)

10.1.2   Gap-Based Coding Strategy and Projection Pattern Design

Our first-version projector used a wave pattern [13]. Because the reflectance conditions inside a human body are very different from ordinary environments, we tailored an original pattern design specifically for the intra-operative environment for the second version. One significant effect under an endoscopic environment is the strong subsurface scattering on the surfaces of internal organs. Although the wave-shaped pattern is sufficiently robust in the normal projector-camera systems of ordinary environments, it is not optimal for the environments wherein the endoscopes were used. If the wave-shaped pattern were to be projected onto surfaces of bio-tissues, some of the important information, including wave curvature would be blurred, because of strong subsurface scattering.

To avoid losing important detailed information, we sought a pattern having a larger, low-frequency structure. Existing patterns of this kind include sparse dots or straight-line-based pattern of wide intervals. However, sparse dots are difficult to decode with wide baselines and large windows, because the patterns are heavily distorted under such conditions. However, a simple line-based pattern cannot efficiently encode distinctive information [14]. Instead, for our second projector, we proposed a line-based pattern with large intervals and a new encoding strategy that was robust to scattering [15]. The pattern consisted of only line segments, as shown in Fig. 10.3b, where the vertical patterned lines were all connected and straight. However, the horizontal segments were designed in a way that left a small variable vertical gap between adjacent horizontal segments along the vertical line. With this configuration, a higher level ternary code emerged from the design, having the following three codewords: S (the end-points of both sides have the same height); L (the end-point of the left side is higher); and R (the end-point of the right side is higher). The codes of the pattern of Fig. 10.3b are shown in color in Fig. 10.3c.

In our system, we assumed a soft limitation on the relative positions between the camera and the projector, for which the epipolar lines, drawn on the pattern image of Fig. 10.3b, did not coincide with the vertical lines. We assumed this limitation, because we used a light-sectioning method from the detected vertical lines. If a vertical line coincided with an epipolar line, reconstruction of the line was not possible. For reconstruction accuracy, directions of the epipolar lines needed to be almost horizontal in the pattern image. However, it was not a strict requirement. Note that, in the proposed pattern, each set of horizontal-line segments at the same column was slightly inclined with angles of specific degrees (e.g., according to a piecewise-long wavelength sinusoid as shown in Fig. 10.3b). Owing to this property, even if the epipolar lines were exactly along the horizontal direction, the number of candidates would be decreased, which is key to decreasing the ambiguity to realize robust and high-precision on shape reconstruction.

10.1.3 Convolutional NN (CNN)-Based Feature Detection and Decoding for Active Stereo

The pattern of the second-version projector had a grid-like structure and discrete codes for each grid point, as shown in Fig. 10.3. To reconstruct the 3D information using the projector-camera system, the pattern was projected onto the target surface, and an image of the projected pattern was captured. From the captured image, the grid structure and the grid point codes were detected. This was not a trivial task, because, in the practical cases wherein endoscopes were used, the target surface comprised bio-tissues having strong subsurface scattering that blurred the projected pattern, and the horizontal edges were not always continuous because of the pattern design.

At first, we used belief-propagation-based line detection to extract the vertical and horizontal lines, and detected the crossing points between them to construct a grid graph aggregating the discontinuous horizontal edges, as shown in Fig. 10.4. We also decoded the codes for each of the grid points [15]. However, this three-step approach, consisting of the line detection, grid construction, and code detection, had a major problem. The accuracies of the grid and code detection were insufficient. One reason for the inaccuracy was the use of sequential-processes error accumulation, wherein errors of line detection disturbed the subsequent grid detection, and the errors of the line/grid detections disturbed the code detection. To address this problem, we developed a new method for detecting the grids and pattern codes from the captured images using U-Nets [16], a variation of a fully convolutional neural network (FCNN).

Detection of Grid Structure

The structure of the U-Net is shown in Fig. 10.5. The numbers in the figure represent dimensions of the feature maps. For example, the single-channel image (i.e., intensity image) of the input was converted to a 64-channel feature map by applying

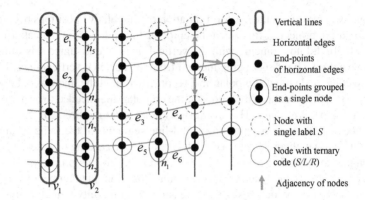

Fig. 10.4 Grid graph structure

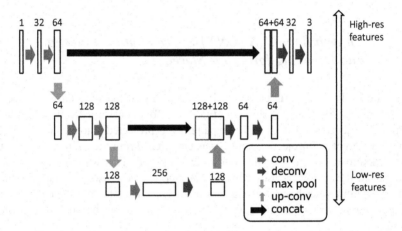

Fig. 10.5 Structure of U-Nets

two convolution steps. Then, the spatial size of the feature map was sub-sampled to become a lower resolution feature map (1/2 for both x- and y-axes) via max pooling. This coarse resolution feature map was later up-sampled via up-convolution and concatenated with the high-resolution feature maps. The information flow in the feature map looks like the "U" character in Fig. 10.5. Thus, it is called a "U-Net". Owing to this network structure, both fine and coarse resolution features are accounted for in the outputs of U-Nets.

U-Nets were originally used for pixel-wise labeling and segmentation of images. Applying a U-Net to an image allows it to finally produce a feature map of the same size. In the resulting N-dimensional feature map, each pixel is an N–D vector. By taking the index of the maximum element for each N–D vector, an image of the N-labels is obtained.

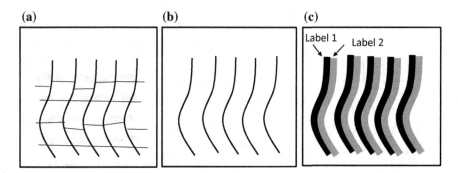

Fig. 10.6 Training data for vertical line detection. **a** Sample image of the projected pattern. **b** Vertical lines that are manually annotated. **c** Labeled regions used as training data

The training process of the U-Net for detecting vertical lines is as follows. First, image samples of the pattern-illuminated scene are collected. Then, the vertical line locations of the image samples are designated manually as curves of one-dot widths. The one-dot-width curves are too sparse and narrow to be directly used as regions of training data. Thus, regions having five-dot widths of the left and right sides of the thin curves are extracted and are labeled as 1 and 2, respectively, as shown in Fig. 10.6. The rest of the pixels are labeled as zero. These three-labeled images are used as training data. Then, a U-Net is trained to produce the labeled regions using the loss function of the softmax entropy between the three-labeled training data and the 3D feature map produced by the trained U-Net.

By applying the trained U-Net to the image, we get the three-labeled image, with which the left and right sides of the vertical curves are labeled as 1 and 2, respectively. Thus, by extracting the two horizontally adjacent pixels in which the left is 1 and the right is 2, and by connecting those pixels vertically, vertical curve detection is achieved.

The horizontal curve detection is similarly achieved. However, the horizontal edges can be disconnected because of the gaps at the grid points. Even in those cases, training data is provided as continuous curves that traverse the center point of the gaps, as shown in Fig. 10.7. By training a U-Net using this training data, we can expect results in which horizontal curves are detected as continuous at grid points, even if they are actually disconnected by gap codes.

An advantage of using U-net for line detection of the grid structure is that the U-Net can be implicitly trained to use not only local intensity variation, but it can also use more global information, such as repetitive information of grid-like structures. Supporting evidence of this improvement can be observed by processing an image sample that is scaled, such that the training image set does not include the similarly scaled images. In this case, line detection performance noticeably worsens, as shown in Fig. 10.8.

The above comparison also shows the importance of providing training datasets that cover possible image variations. This may seem to be a problem with the pro-

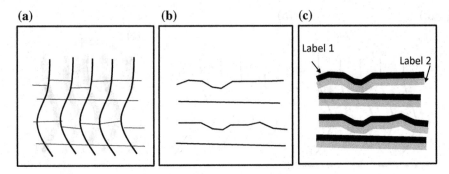

Fig. 10.7 Training data for horizontal-line detection. **a** sample image of the projected pattern. **b** Horizontal lines that are manually annotated. **c** Labeled regions used as training data. Note that the discontinuity at the grid points is intentionally connected in the teacher data

posed method, because we should provide sufficiently annotated training samples. However, this limitation is significantly relaxed by the fact that the training set can be considerably smaller compared with other deep-learning applications, such as natural-image classification. This is because, in active-stereo methods, we can control the pattern projection setup. Thus, the variation of the projected pattern is more limited than with natural-image classification.

Another advantage of using the U-Net for grid detection is that the horizontal edges that are actually disconnected by the gaps are intentionally detected as continuous curves by providing training data. This task is not simple for conventional line detection algorithms. Thanks to the continuously detected horizontal curves, analysis of grid structure becomes much simpler than in previous works [15].

The max pooling and up-convolution of the U-Net provides feature maps for different resolutions. For line detection, we use four different resolutions. In the coarsest resolutions, the size of a pixel feature map is 8×8 pixels times the original image. Thus, the convolution in this resolution uses the information of about 24×24-pixel patches, which is larger than typical grid size of about 20×20. Thus, the U-Net uses information about the grid structures for local line detections.

Detection of Pattern Codes

In the proposed method, the identification of gap codes is processed by directly applying the U-Net to the image signal, not from the line detection results. Thus, the gap-code estimation does not depend on line-segment detection, which is advantageous for stable detection of gap codes. Note that this direct method is not easy to implement via conventional image processing.

The training data generation is shown in Fig. 10.9. During the training process, the white background pixels of Fig. 10.9c are treated as "don't care" regions. The advantage of directly detecting a pattern code is the stability of code detection. In a previous work [15], identification of gap codes was achieved by using results of line detection, failure of line detection, and failure of grid-structure analysis, which

(a) (b) (c) (d)

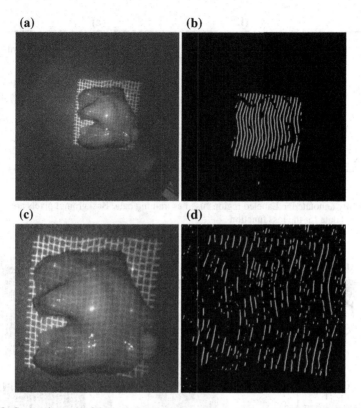

Fig. 10.8 Scale mismatch of the input image. **a** Input image with matched scale. **b** CNN (vertical line detection) result of (**a**). **c** Input image with mismatched scale (doubled in the x- and y-axes). **d** CNN (vertical line detection) result of (**c**)

leads to code detection failures. The proposed method was free from such problems of sequential processing.

10.1.4 Self-calibration of Camera and Projector Relative Pose

In this system, the pattern projector was not fixed to the endoscope as shown in Fig. 10.10a. Thus, pre-calibration of extrinsic parameters was not possible. To deal with this problem, we proposed an autocalibration method wherein frame-wise calibration was used [9]. For autocalibration of the pattern projector's position and orientation, we added nine bright markers to the projected pattern shown as white dots in Fig. 10.10b. We detected the positions and IDs of the imaged markers using specially trained fully convolutional networks (FCNs). The detected marker positions and IDs were prone to errors. We filtered out the erroneously detected markers

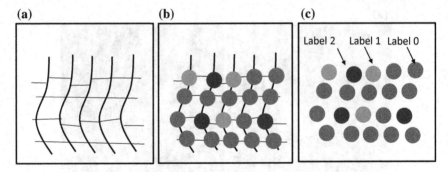

Fig. 10.9 Training data for code detection. **a** Sample image of the projected pattern. **b** Codes that are manually annotated. **c** Labeled regions used as training data. Background pixels are treated as "don't care" data for the loss function

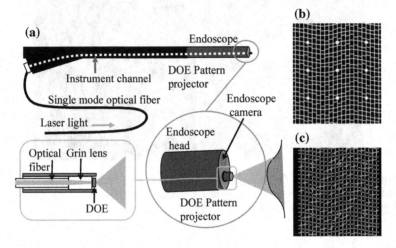

Fig. 10.10 DOE-based pattern projector for endoscope. **a** System configuration of the 3D endoscopic system. **b** The projected pattern with nine bright markers and gap coding. **c** The code embedded as gaps at grid points of the projected pattern

by removing markers that could not maintain stable positions and IDs for a certain period of time in the image sequences for five frames. Because the projector was not fixed to the endoscope, the extrinsic parameters changed while scanning. However, the change rate was assumed to be relatively slow. Thus, we estimated the projector's extrinsic parameters by using a certain number of frames around each. For example, we could use 90–110 marker frames to estimate the extrinsic for frame 100. Otherwise, we could use 91–111 marker frames to estimate the extrinsic for frame 101, etc. Figure 10.11 shows the processes of the self-calibration, where an original image, results of marker detections, and estimated and visualized epipolor lines are shown.

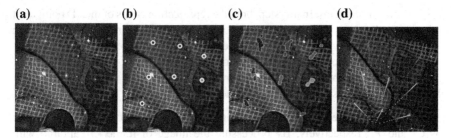

Fig. 10.11 A self-calibration process. **a** A source image. **b** Detected marker positions with their color-represented IDs. **c** Marker positions for multiple frames. **d** Calibration results shown as nine epipolar lines of the markers

10.1.5   Multi-view Measurement Integration

To address problems associated with limited reconstructed regions, we proposed an extended bundle-adjustment (BA) technique that integrates multiple shapes into a consistent single shape by simultaneously estimating 3D shapes and the calibration parameters. If shapes are distorted because of calibration errors, they cannot be registered correctly using extant techniques.

As discussed, the pattern projector was not fixed to the endoscope, and the projector's pose parameters varied frame by frame. Thus, the parameters had to be autocalibrated for each frame. However, because estimations could include errors that cause shape distortions, shapes from multiple frames could not be registered using a simple iterative closest-point (ICP) algorithm. Additionally, a naive BA for multi-frame data was not applicable, because the sequence of active-stereo data only included correspondences between the projector and the camera within each frame instead of correspondences between different frames. Consequently, we proposed an active BA algorithm, where an ICP-like inter-frame correspondence search and a BA were applied alternatively. The steps of the active BA algorithm are as follows.

Step 1.   Initial parameters of the relative pose between the projector and the camera and the positions of the shapes of the frames are given as input.
Step 2.   The 3D shape for each frame is reconstructed from the current pose information and the given correspondences between the projector and the camera for each frame.
Step 3.   The corresponding points between different frames are sampled using a proximity relation between the frame surfaces. This is similar to the process used to retrieve the corresponding point pairs in an ICP algorithm.
Step 4.   Using the inter-frame correspondences obtained in Step 3, corresponding pairs of 3D points with associated 2D projection information onto the projectors and the cameras are generated. This step will be explained later in more detail.
Step 5.   Locations of the 3D points with 2D projection information obtained in Step 4 are optimized with the intrinsic and extrinsic parameters of the projectors

and the cameras. In this step, both the reprojection errors of the 3D points and the distances between the corresponding points between different frames are minimized with respect to the pose and intrinsic parameters of the projectors and the cameras. Using the solutions of minimization, dense 3D shapes and the relative position between frames are updated.

Step 6. Repeat Steps 2–5 until convergence.

Step 4 is illustrated in Fig. 10.12. For frame k, denote the camera and projector as C_k and P_k, respectively. By processing the image of frame k (captured with C_k), the 2D correspondences between C_k and P_k are obtained. Let the jth pair of correspondences be a pair, $\mathbf{u}_{k,j}^{c}$ of C_k and $\mathbf{u}_{k,j}^{p}$ of P_k. The 3D point obtained by the triangulation of $\mathbf{u}_{k,j}^{c}$ of C_k and $\mathbf{u}_{k,j}^{p}$ of P_k is denoted $\mathbf{p}_{k,j}$. Let frame l be another frame. Furthermore, all correspondence pairs of frame l are reconstructed. If the reconstructed points, $\mathbf{p}_{l,j}$, are sparse, they should be interpolated, and the dense depth image, D_l, with the view, C_l, is obtained. Furthermore, $\mathbf{p}_{k,j}$ is projected onto D_l using the pose and intrinsic parameter of C_l. If the projected pixel is a valid 3D point, we can define this point in the corresponding point, $\mathbf{p}_{k,j}$, in frame l. Let this corresponding point be $\pi_l(\mathbf{p}_{k,j})$. The 2D projection of $\pi_l(\mathbf{p}_{k,j})$ can be calculated using camera C_l and projector P_l. Let these 2D points be $\mathbf{v}_{k,j,l}^{c}$ and $\mathbf{v}_{k,j,l}^{p}$, respectively. $\mathbf{p}_{k,j}$ and $\pi_l(\mathbf{p}_{k,j}^{c})$ are corresponding points between different frames. Generally, these corresponding points differ, but they are neighboring points of frames k and j.

In our algorithm, we calculated BA-style reprojection errors of points $\mathbf{p}_{k,j}$ and $\pi_l(\mathbf{p}_{k,j}^{c})$, Respectively, within each frame k and l, and the distance errors between the corresponding points. Then, the total cost to be minimized was the weighted sum of reprojection errors of all points $\mathbf{p}_{k,j}$ and distance errors of all pairs of $\mathbf{p}_{k,j}$ and $\pi_l(\mathbf{p}_{k,j}^{c})$.

The cost function to be minimized is expressed as follows:

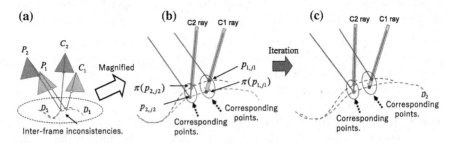

Fig. 10.12 Active BA principle. **a** Correspondence finding process and cost between points. **b** An ICP-like correspondence search between frames. **c** BA-like optimization of camera and projector parameters for all frames were applied iteratively until convergence

$$L(I, E, P) = \sum_{k} \sum_{j} \{\text{reproj}(\mathbf{p_{k,j}}; I_{C_k}, E_{C_k})$$

$$+ \text{reproj}(\pi_l(\mathbf{p_{k,j}^c}); I_{P_k}, E_{P_k})\}$$

$$+ w_c|\mathbf{p_{k,j}} - \pi_l(\mathbf{p_{k,j}^c})|^2 + w_b\{S(E) - \text{Const}\}^2, \qquad (10.1)$$

where I_{C_k} and E_{C_k} are intrinsic and extrinsic parameters of camera C_k, I_{P_k}, and E_{P_k} are intrinsic and extrinsic parameters of projector P_k, reproj() denotes the BA-style reprojection errors, I is the set of intrinsic parameters I_{C_k} and I_{P_k}, E is the set of extrinsic parameters, E_{C_k} and E_{P_k}, and P is the set of $\mathbf{p_{k,j}}$ and $\pi_l(\mathbf{p_{k,j}^c})$. $L(I, E, P)$ is minimized with respect to I, E, and P. $S(E)$ is a scale function that determines the scale of the scene, and Const is a constant value. We use the sum of distances for randomly sampled devices (e.g., projectors or cameras) for $S(E)$. w_c and w_b are weights for the cost terms. In real applications, such as the endoscopic systems considered in this study, the intrinsic parameters of the camera for all frames, k, are the same. Thus, we use common intrinsic for all k.

The active BA can be considered a variation of the ICP algorithm, where ICP is used to estimate only a rigid transformation between frames. However, in this study, the proposed algorithm estimates the projector-camera relative pose, which significantly affects the shapes of the frames.

10.1.6 Evaluation and Experimental Results

Gap-Coding Experiments

To evaluate the system under realistic conditions, a biological specimen extracted from the stomach of a human via endoscopy operation was measured, as shown in Fig. 10.13. We captured the tissue from different distances with about 25 and 15 mm. Using the proposed algorithm, both images were nearly fully reconstructed, except for the regions that were affected by the bright specular highlights. Avoiding the effects of these specular is future work. Other than these points, the correspondences between the detected grid points and the projected pattern were accurately estimated, and the shapes of the specimen were successfully reconstructed.

Evaluation of CNN-Based Robust Grid/code Detection

To evaluate the proposed CNN-based pattern-feature extraction for endoscopic images, we measured resected cancer specimens. The dataset used to train the CNN was obtained from 47 actual endoscopic images in which the pattern projected by the DOE pattern projector was as shown in Fig. 10.10b and c. The training dataset was annotated as shown in Fig. 10.6. From this set, sub-image patches (120×120) were extracted and trained for 4,000 iterations using Adam. For each iteration, 20 sub-images having rotation augmentation ($-30°$ to $30°$) and scaling (0.5:2) were input as a batch.

The appearance of the specimen, the image captured by the 3D endoscope with the projected pattern, and the U-Net outputs for horizontal-line detection and code-

(a)   **(b)**   **(c)**

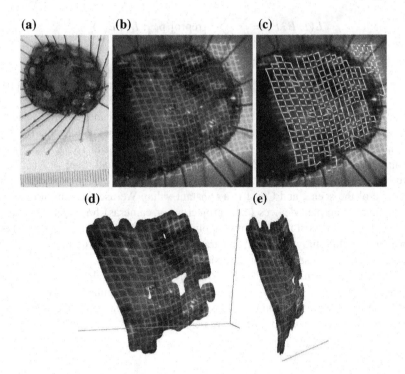

(d)   **(e)**

Fig. 10.13 3D reconstruction of real tissues from a human stomach. **a** The appearance of the sample with a ruler with marks with 1-mm intervals. **b** The captured image with distance of about 25 mm. **c** The detected grid graph with gap codes. **d** The reconstructed shape seen from the front. **e** The reconstructed shape seen from the side

letter detection are shown in Fig. 10.14a, b, c, and d, respectively. The grid structures and codes extracted from the U-Net result are shown in Fig. 10.14e and the grid extracted using a previous technique [15] is shown in Fig. 10.14f. By comparing the code colors between Fig. 10.14e and f, we can confirm that, in terms of code extraction, the proposed method returned better results than the earlier techniques.

We applied the proposed and earlier techniques to other specimens. The extraction results are summarized in Fig. 10.14g. The graph represents the numbers of grid points where gap codes were correctly or erroneously estimated. As can be seen, the total number of points detected using the previous technique [15] was larger than the number detected using the proposed technique. However, a large number of error points were included in the total number of detected points. Therefore, the total number of correctly detected points was greater with the proposed technique for all the cases.

Compared to the conventional approach [15], the proposed approach was more stable because of the *contextual* line detection ability of a CNN. In other words, continuous lines were detected, taking the grid-like structures around the local area into account, even if the local pixel intensity signals were extremely weak. Additionally,

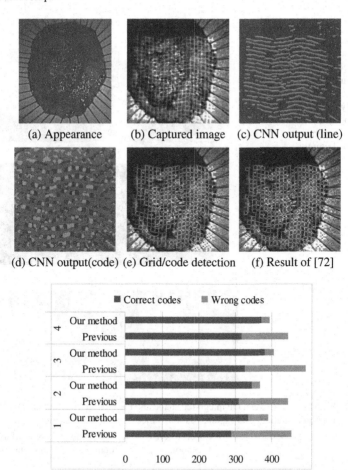

(a) Appearance   (b) Captured image   (c) CNN output (line)

(d) CNN output(code) (e) Grid/code detection   (f) Result of [72]

(g) Numbers of correct/wrong detections of
gap codes for 4 polyp specimens

Fig. 10.14 Grid and code detection results for a cancer specimen: **a** Appearance. **b** Captured image. **c** U-Net output for horizontal-line detection. **d** U-Net code detection output. **e** Extracted grid structures and codes using the proposed method. **f** Extracted grid structures and codes using a previous method [15]. **g** Number of correct and incorrect detections of gap codes for four polyp specimens

the ability to detect codes directly from input images contributes to the stability of the proposed approach. Even if a detection failure were to occur, such as a continuous line detected as separate lines, such failures would not affect code detection. However, with the conventional approach [15], code detection was affected by line detection.

Fig. 10.15 Entire shape of liver phantom reconstruction result. **a** Captured image for active stereo. **b** Ground-truth shape captured using Gray code projection. **c** Camera positions (yellow/cyan), projector positions (orange/azure), and reconstructed points (red/blue) before/after BA. **d** 3D points before BA. **e** 3D points after BA (Color figure online)

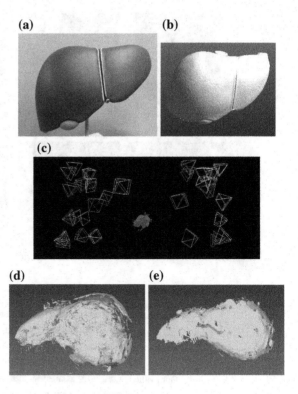

(a) (b) (c) (d) (e)

Table 10.1 RMSE[mm] between the integrated shape and the ground truth obtained by KinectFusion and the proposed method. The captured sample object was the same as Fig. 10.15

Initial	Proposed	KinectFusion (1 rotation)	KinectFusion (2 rotations)	KinectFusion (3 rotations)
8.37	2.49	4.11	3.90	3.13

Evaluation of Active BA Using CCD-Based System

In an experiment to evaluate active BA, we initially, captured a single object (i.e., a liver phantom) multiple times (15 sets) using an experimental projector-camera system, equipped with a CCD camera. The liver phantom was approximately the same size as a human liver. We used a CCD-camera system instead of an endoscopic system, because we wanted to capture precise shape data that could be used as ground truth simply by projecting Gray code patterns [17] using an off-the-shelf video projector. Obtaining effective ground-truth shapes is difficult with endoscopic cameras because of their low resolution and inherent fish-eye distortions. In our evaluation, a grid pattern was projected onto the object (Fig. 10.15a), and the images were captured by a CCD camera. The relative positions of the projector and the camera differed slightly for each frame to simulate the endoscopic system. Under these conditions, autocalibration and 3D reconstruction processes were performed for each frame. The ICP alignment results for the initial shapes and the active BA results based on these data are shown in Fig. 10.15e. The alignment results for the initial shape differed considerably between frames. However, the differences were reduced significantly by active BA. We also scanned the same object using KinectFusion [18] with the camera rotating around the target 1, 2, and 3 times. The results are shown in Fig. 10.15c, d, and e, respectively. KinectFusion does not have a loop-closure mechanism. Consequently, inconsistent shapes were reconstructed. To evaluate the accuracy of the results, we used a commercial scanner to capture the same object as the ground truth, as shown in Fig. 10.15b. Then, we applied ICP to calculate RMSEs (Table 10.1). The results indicate that owing to simultaneous optimization of entire shapes and parameters, the proposed technique could recover consistent shapes more accurately than KinectFusion.

Scan Using a Pig's Stomach

To confirm the strength of the proposed method for actual operations, we scanned a pig's stomach (Fig. 10.16a). Because the projector and camera could not be rigidly attached, the relative position of the devices changed continuously during the operation. We applied the proposed autocalibration technique to retrieve the initial pose and shape for each frame, and we then applied active BA. Figure 10.16c and d are images captured under normal and structured light conditions, respectively. Figure 10.16e is the extracted grids and codes. The merged shape after active BA is superimposed on Fig. 10.16f. Figure 10.16g–i are images processed in the same manner as for another frame. Figure 10.16k–m are shapes colored by frames prior to active BA and after active BA (210 iterations). Note that the gaps between shapes were minimized as active BA iterations procedure, as shown by the red ellipses in the images. Figure 10.16n and o are merged shapes after active BA. In Fig. 10.16n, the regions outlined in red and blue are the regions captured in Fig. 10.16c and g, respectively. From Fig. 10.16d, j, n, and o, we can confirm that a single, consistent merged shape was obtained.

10.2 Heart Rate Measurement Using a High-Speed Active-Stereo 3D Scanner

10.2.1 Noncontact Heart Rate Measurement Techniques

Monitoring vital signs, such as heartbeat or respiration, is an important technique for health care or sport-training applications. In the conventional method of measuring the heartbeat, electrodes are put on the skin and the electrocardiogram (ECG) is measured. This method can measure the heartbeat accurately, however it requires direct contact with the skin which often causes discomfort and stress. In addition, it requires certain preparation and takes time, noncontact measurement is desired, especially for daily monitoring of severe patient or aged persons. Using active light-

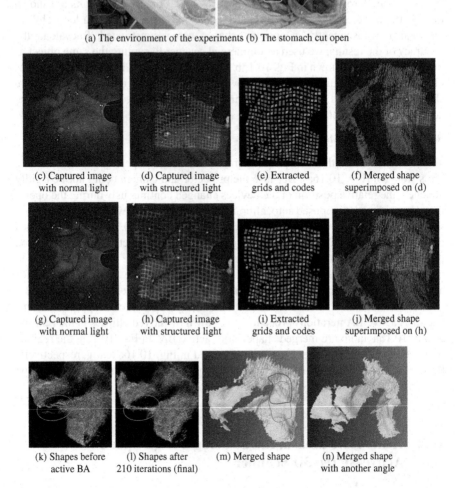

(a) The environment of the experiments (b) The stomach cut open

(c) Captured image with normal light

(d) Captured image with structured light

(e) Extracted grids and codes

(f) Merged shape superimposed on (d)

(g) Captured image with normal light

(h) Captured image with structured light

(i) Extracted grids and codes

(j) Merged shape superimposed on (h)

(k) Shapes before active BA

(l) Shapes after 210 iterations (final)

(m) Merged shape

(n) Merged shape with another angle

Fig. 10.16 3D reconstruction of bio-tissue inside a pig stomach. **a** Experimental environment. **b** Pig-stomach cut open after the experiment. The yellow rectangle is the area scanned by a single frame; the blue region is the integrated shape region. **c** The appearance inside the stomach with marker positions. **d** The captured image with the pattern projected. **e** Extracted grids and codes. **f** Merged shape after active BA superimposed on (**d**). **g**–**j** Images processed in the same manner as (**c**)–(**f**) for another frame. **k**, **l** Shapes colored by frames before and after active BA. **m** and **n** Merged shape after active BA. In (**m**), the regions outlined in red and blue are the regions captured in (**c**) and (**g**), respectively (Color figure online)

ing and cameras for such measurements can provide a better solution, because it achieves noncontact measurement with relatively high accuracy. There are several noncontact techniques. Among the noncontact heartbeat measurements, imaging photoplethysmography (iPPG) has received much attention. iPPG is the technique of measuring the pulse waveform based on the optical properties of the human skin. The principle is that the light absorption in skin changes with changing concentration of hemoglobin in the blood with each heartbeat [19, 20]. In these methods, the heartbeat is usually estimated by measuring the changing intensities on the face. However, it is difficult to track the face accurately, especially in the presence of motion, and thus, those image-based techniques are weak with subject motion and environmental illumination changes.

10.2.2   Grid Pattern Based Heart Rate Measurement System

To overcome this problem, a noncontact measurement of cardiac beat that uses an active-stereo technique, leveraging the wave-shaped grid projection system was proposed by Sagawa et al. [13]. In the method, waved-grid pattern was projected from a video projector to the subject's breast, as shown in Fig. 10.17. To capture the high-speed motion of the subject's breast, a high-speed camera that was capable of capturing images at 100 FPS was used. The heartbeat was extracted as the small motion of the surface depth for the area of the subject's breast. To extract these signals, Fourier analysis was used to separate the motion into high-band-pass signals (Fig. 10.18). For a cardiac heartbeat, 0.4–5-Hz signals were extracted and for respiration motions, signals under 0.4 Hz were extracted.

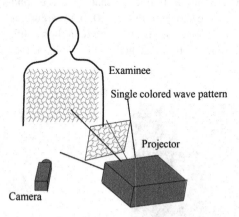

Examinee

Single colored wave pattern

Projector

Camera

Fig. 10.17 Heartbeat capturing system. Cited from [21] with agreement to the authors

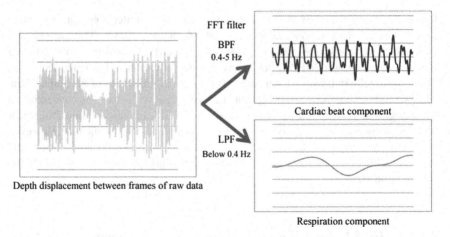

FFT filter

BPF
0.4-5 Hz

Cardiac beat component

LPF
Below 0.4 Hz

Depth displacement between frames of raw data

Respiration component

Fig. 10.18 Extraction of vital signs. Cited from [21] with agreement to the authors

10.2.3 Evaluation and Experimental Results

Actual measurements by an experimental system are executed to examine the validity
of our proposed method. In the experiment, SILICON VIDEO monochrome 643M
(EPIX Inc.) at a frame rate of 100 FPS with VGA resolution and the video projector
EB-1750 (EPSON Corporation) were used. The distance between the camera lens
and the projector lens was set to 600 mm. The motion of the breast motion of the
examinees who were wearing white t-shirts were examined. Figure 10.19 shows an
example of the captured image and the reconstructed 3D shape of the image with an
x-shaped marker where motion frequency analyses were conducted. The results of
the frequency analyses are shown in Fig. 10.20. In the figure, measured results from
a wireless ECG are also shown as ground-truth data. It is confirmed that the timings
of the extracted heartbeat motion matched to the results of the ECG.

(a)   **(b)**

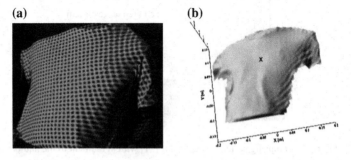

Fig. 10.19 **a** Input image. **b** Reconstructed 3D shape with an x-shaped marker showing the position
of the motion analysis. Cited from [21] with agreement to the authors

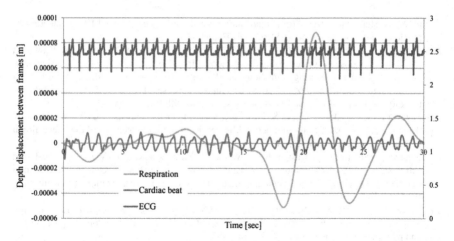

Fig. 10.20 Measurement results compared with ECG as the ground truth. Cited from [21] with agreement to the authors

References

1. Okatani T, Deguchi K (1997) Shape reconstruction from an endoscope image by shape from shading technique for a point light source at the projection center. Comput Vis Image Underst J (CVIU) 66(2):119–131
2. Wu C, Narasimhan SG, Jaramaz B (2010) A multi-image shape-from-shading framework for near-lighting perspective endoscopes. Int J Comput Vis (IJCV) 86(2–3):211–228
3. Ciuti G, Visentini-Scarzanella M, Dore A, Menciassi A, Dario P, Yang GZ (2012) Intra-operative monocular 3d reconstruction for image-guided navigation in active locomotion capsule endoscopy. In: 2012 4th IEEE RAS & EMBS international conference on biomedical robotics and biomechatronics (BioRob). IEEE, pp 768–774
4. Stoyanov D, Scarzanella MV, Pratt P, Yang GZ (2010) Real-time stereo reconstruction in robotically assisted minimally invasive surgery. In: International conference on medical image computing and computer-assisted intervention. Springer, pp 275–282
5. Visentini-Scarzanella M, Hanayama T, Masutani R, Yoshida S, Kominami Y, Sanomura Y, Tanaka S, Furukawa R, Kawasaki H (2015) Tissue shape acquisition with a hybrid structured light and photometric stereo endoscopic system. In: Computer-assisted and robotic endoscopy. Springer, pp 46–58
6. Grasa OG, Bernal E, Casado S, Gil I, Montiel J (2013) Visual slam for handheld monocular endoscope. IEEE Trans Med Imaging 33(1):135–146
7. Aoki H, Furukawa R, Aoyama M, Hiura S, Asada N, Sagawa R, Kawasaki H, Tanaka S, Yoshida S, Sanomura Y (2013) Proposal on 3-d endoscope by using grid-based active stereo. In: 2013 35th annual international conference of the IEEE engineering in medicine and biology society (EMBC). IEEE, pp 5694–5697
8. Furukawa R, Aoyama M, Hiura S, Aoki H, Kominami Y, Sanomura Y, Yoshida S, Tanaka S, Sagawa R, Kawasaki H (2014) Calibration of a 3d endoscopic system based on active stereo method for shape measurement of biological tissues and specimen. In: 2014 36th annual international conference of the IEEE engineering in medicine and biology society. IEEE, pp 4991–4994
9. Furukawa R, Masutani R, Miyazaki D, Baba M, Hiura S, Visentini-Scarzanella M, Morinaga H, Kawasaki H, Sagawa R (2015) 2-dof auto-calibration for a 3d endoscope system based

on active stereo. In: 2015 37th annual international conference of the IEEE engineering in medicine and biology society (EMBC). IEEE, pp 7937–7941

10. Furukawa R, Sanomura Y, Tanaka S, Yoshida S, Sagawa R, Visentini-Scarzanella M, Kawasaki H (2016) 3d endoscope system using doe projector. In: 2016 38th annual international conference of the IEEE engineering in medicine and biology society (EMBC). IEEE, pp 2091–2094

11. Furukawa R, Naito M, Miyazaki D, Baba M, Hiura S, Kawasaki H (2017) HDR image synthesis technique for active stereo 3d endoscope system. The 39th EMBC, pp 1–4

12. Sagawa R, Kawasaki H, Kiyota S, Furukawa R (2011) Dense one-shot 3d reconstruction by detecting continuous regions with parallel line projection. In: Proceedings of the international conference on computer vision (ICCV). IEEE, pp 1911–1918

13. Sagawa R, Sakashita K, Kasuya N, Kawasaki H, Furukawa R, Yagi Y (2012) Grid-based active stereo with single-colored wave pattern for dense one-shot 3d scan. In: 2012 second international conference on 3D imaging, modeling, processing, visualization & transmission. IEEE, pp 363–370

14. Kawasaki H, Furukawa R, Sagawa R, Yagi Y (2008) Dynamic scene shape reconstruction using a single structured light pattern. In: Proceedings of the IEEE conference on computer vision and pattern recognition (CVPR). IEEE, pp 1–8

15. Furukawa R, Morinaga H, Sanomura Y, Tanaka S, Yoshida S, Kawasaki H (2016) Shape acquisition and registration for 3d endoscope based on grid pattern projection. In: Proceedings of the European conference on computer vision (ECCV). Springer, pp 399–415

16. Ronneberger O, Fischer P, Brox T (2015) U-net: convolutional networks for biomedical image segmentation. In: International conference on medical image computing and computer-assisted intervention. Springer, pp 234–241

17. Inokuchi S (1984) Range imaging system for 3-d object recognition. In: Proceedings of the international conference on pattern recognition (ICPR), pp 806–808

18. Newcombe RA, Izadi S, Hilliges O, Molyneaux D, Kim D, Davison AJ, Kohi P, Shotton J, Hodges S, Fitzgibbon A (2011) Kinectfusion: real-time dense surface mapping and tracking. In: 2011 10th IEEE international symposium on mixed and augmented reality. IEEE, pp 127–136

19. Poh MZ, McDuff DJ, Picard RW (2010) Advancements in noncontact, multiparameter physiological measurements using a webcam. IEEE Trans Biomed Eng 58(1):7–11

20. Rahman H, Ahmed MU, Begum S, Funk P (2016) Real time heart rate monitoring from facial RGB color video using webcam. In: The 29th annual workshop of the swedish artificial intelligence society. http://www.es.mdh.se/publications/4354-

21. Aoki H, Furukawa R, Aoyama M, Hiura S, Asada N, Sagawa R, Kawasaki H, Shiga T, Suzuki A (2013) Noncontact measurement of cardiac beat by using active stereo with waved-grid pattern projection. In: 2013 35th annual international conference of the IEEE engineering in medicine and biology society (EMBC). IEEE, pp 1756–1759

Chapter 11
E-Heritage

Cultural heritage is a valuable asset left to us by our ancestors. Unfortunately, cultural heritage is routinely lost via natural disasters (e.g., earthquakes and tsunami) and man-made disasters (e.g., arson and mischief). Physical preservation efforts have been made to physically protect heritage artifacts. Some artifacts have thus been moved to museums and made available to the public. Others have been removed from public view or left to be closed at their original excavation sites. In parallel to these physical efforts, new digital preservation efforts have been made to store objects in virtual spaces by using recent advanced sensing technologies and active-lighting techniques. We have carried out digital preservation projects that recorded geometric and photometric information of various heritage sites using active-lighting sensors. The collection of digitized data is referred to as e-heritage.

Typically, e-heritage projects require measurements of outdoor assets. However, owing to severe illumination changes and shadow effects caused by weather changes, active-lighting sensors possessing their own light sources are advantageous. Additionally, these active-lighting sensors are guaranteed to have the accuracy in measurement results. Therefore, with the e-heritage projects described in this chapter, all measurements were performed using active lighting.

This chapter introduces two representative e-heritage projects conducted by us. One is the Kyushu tumuli project [1] and the other is the Cambodian project [2]. The Kyushu tumuli project and the Cambodian project are presented as representative examples to record photometric and geometric information, respectively. Both projects are commonplace in the sense that their aims were to record original heritage information by using active-lighting sensors. However, their emphases were slightly different from one another. The Kyushu tumuli project emphasized acquisition methods of precise photometric information and analysis methods to obtain new findings based on the photometric information. For the Cambodian project, owing to the size of the temples involved, our primal goal was to develop active-lighting sensors to quickly obtain geometric shape information. We also considered the interpretations of these shapes in terms of archeology.

© Springer Nature Switzerland AG 2020   263
K. Ikeuchi et al., *Active Lighting and Its Application for Computer Vision*,
Advances in Computer Vision and Pattern Recognition,
https://doi.org/10.1007/978-3-030-56577-0_11

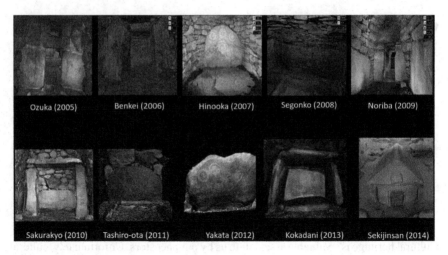

Fig. 11.1 Kyushu painted tumuli

11.1 Photometric Aspect: Kyushu Decorative Tumuli Project

Many decorative tumuli with mural paintings constructed around the sixth century exist on Kyushu Island, Japan. Most of these tumuli are closed to public to protect their colors from CO_2 emissions and other harm. Due to the low profile of the tumuli, the Kyushu National Museum decided to make video contents of these mural paintings. Among those tumuli currently existing in Kyushu Island, we have selected ten of them, shown in Fig. 11.1, as representative ones and collected their 3D geometric and photometric data. We used active-lighting sensors inside the tumuli, where it was otherwise completely dark, with spectral cameras and a high color-rendering light source that reproduced sunlight for the photometric data. We also use the Lidar active-lighting sensor to obtain 3D geometric data.

Those obtained data were not only used for video-contents creation but also for archeological analysis, which we refer to as "Cyber archeology." This section explains what kind of active-lighting sensors were used for this purpose and what kind of new findings were obtained along the line of *cyber archeology*.

11.1.1 Ozuka Tumulus

The Ozuka Tumulus is a large keyhole-shaped tomb mound, built in the middle of the sixth century, located in Keisen-cho, Fukuoka Prefecture, as shown in Fig. 11.2. Mural paintings, including both concrete patterns of horses, and abstract patterns (e.g., triangular and bi-legged ring-shapes) almost entirely cover the interior walls

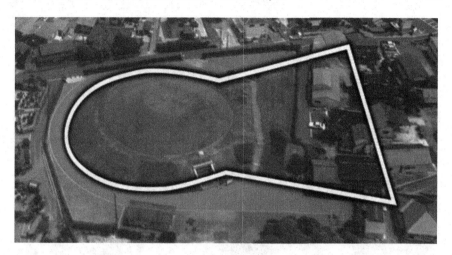

Fig. 11.2 Ozuka Tumulus

of the horizontal stone chamber. The Ozuka Tumulus is a historical treasure in terms of color use, including reds, yellows, whites, blacks, greens, and grays. It is the only known tumulus in which all six colors are used.

Data Acquisition for Video-Contents Creation

We obtained 3D geometric data in a point-cloud format by using an active-lighting Lidar sensor: the Z+F imager. This commercially available active-lighting sensor projects amplitude-modulated (AM) laser light onto objects. It is used to determine the distance to an object by analyzing the differences between phases of the outgoing and returned light. To allow omnidirectional measurement, the sensor applies the laser light to a rotating polygon mirror to sweep one vertical plane. Then, it changes the axis of rotation to sweep the horizontal directions.

Figure 11.3a shows the scene of data acquisition. We scanned the entire mound as well as the inside of the stone chamber. Then, we combine those partial data into the entire 3D geometric data of the tumulus as shown in Fig. 11.3b [3]. For this process, we used the simultaneous alignment algorithm explained later in the Bayon section.

Mapping color images to the entire 3D geometric data requires us to take pictures without leaving gaps. Taking pictures randomly and manually stitching them together gives unsatisfactory results, because of differences in resolution and vignetting. Thus, we mounted a color camera and a high color-rendering light source to an omnidirectional automatic tripod head. The light source was mounted as near as possible to the viewpoint of the camera to avoid self-shadow regions. This system allowed us to quickly obtain omnidirectional high-resolution color images. Figure 11.3c, d show the texture-mapped result to the 3D geometric data by using the omnidirectional color images taken with this device. This data was used to create video contents of the permanent exhibit at the Kyushu National Museum.

(a)   **(b)**

(c)   **(d)**

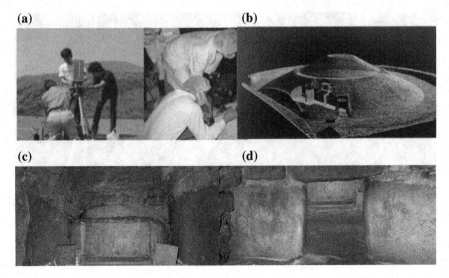

Fig. 11.3 Ozuka tumulus. **a** Measuring the tumulus with a laser range sensor. **b** Obtained point-cloud data of the round mound and the stone chambers. **c** Texture-mapped result of burial chamber. **d** Texture-mapped result of front chamber

Line Spectroscope

We developed a line-filter-based spectroscopy system [4] with which 3D texture-mapped was used not only for generating video contents, but also for archeological analysis. However, to conduct detailed analyses of the coloring pigments, RGB was insufficient. Thus, it was necessary to examine the spectral distribution via spectroscopy. The observed brightness at each wavelength λ is described as

$$L_{observe}(\lambda) = L_{source}(\lambda)S(\lambda), \tag{11.1}$$

where $L_{observe}$, L_{source}, and S represent the observed brightness, the brightness of the illuminating light, and the reflectance ratio of the surface, respectively. The digital color camera obtained pixel values using RGB filters.

$$I_k = \int L_{source}(\lambda)S(\lambda)R_k(\lambda)d\lambda, \tag{11.2}$$

where k, I_k, and R_k denote each color channel, RGB, the observed pixel value, and the spectral response for each channel, respectively. This formula tells us two facts. The first is that the observed RGB value, I_k, depends on the illumination spectral distribution, L_{source}. The second is that, because the camera response curve, R_k has a wavelength range and it is not a delta function of a particular wavelength, the original reflectance, S, cannot be obtained by simple division of observed RGB values by illumination RGB values.

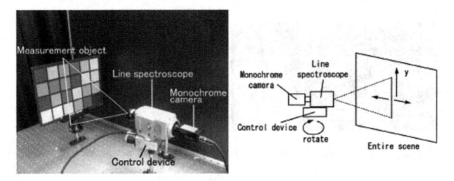

Fig. 11.4 A line spectroscope mounted in front of a monochrome camera

We mount a spectral line filter in front of a monochrome camera as shown in Fig. 11.4. Using the configuration depicted in the figure, the filter has a prism mounted in the vertical direction, and the input vertical light was expanded in the horizontal direction according to their wavelengths. The image, $I(w, y)$, obtained by the camera, depicts the spectral power distribution at wavelength, w, in the horizontal direction and at the y pixel position in the vertical direction.

Putting this system on a rotating table enables us to scan the entire scene. The system measured the spectral distribution of the entire scene by taking multiple images while rotating about the vertical axis.

Cyber Archeology: Painting Condition

One archeological question which we address is under what lighting condition were the color patterns drawn. When the tumulus is completed, the inside is pitch black. Thus, the ancient painter could not have drawn the patterns without illumination. One possibility is that the artist used torchlight *after completion*. Another possibility is to draw them under sun *before construction was complete* and then close the stone chamber.

We obtained the spectral distribution of torchlight and sunlight as shown in Fig. 11.5. Figure 11.6a shows the mural's appearance in sunlight, while Fig. 11.6b shows it under torchlight [3]. Part A indicates the area painted with green clay and Part B indicates the area painted with Manganese clay. These green and manganese clay regions formed a repeating pattern of triangular shapes. In sunlight, the two regions can be distinguished, whereas in torchlight, it is quite difficult to distinguish them. This indicates that when the ancient painter painted the patterns, he/she most likely worked under sunlight condition. Thus, we can conjecture that the patterns were more likely painted prior to the completion of construction while the area was illuminated by the sun. This is a great example of e-heritage data analysis. In this case, spectral images obtained by an active-lighting sensor provided a novel interpretation of how the tumulus was constructed.

(a)   (b)

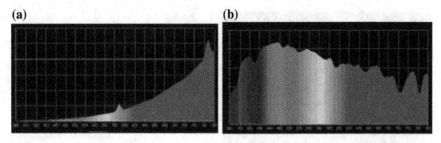

Fig. 11.5 Spectral distribution of torch light and sun light. **a** Torchlight. **b** Sunlight

(a)   (b)

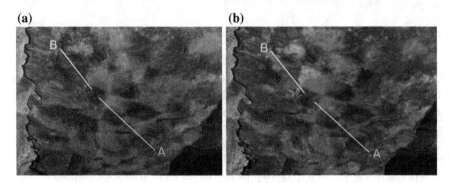

Fig. 11.6 Simulation results. Some regions cannot be distinguished under torchlight, indicating the high possibility that these patterns were drawn in the middle of construction. **a** Appearance under sunlight. **b** Appearance under torchlight

11.1.2 Noriba Tumulus

The Noriba tumulus is also a keyhole-shaped tomb, built in the second half of the sixth century, located in Yame-shi, Fukuoka Prefecture. Concentric circles, triangles, swords, and tough patterns exist in the interior of the stone chamber. Unfortunately, the condition of the mural paintings is not good, and the presence of deposits on the paintings' surfaces makes it difficult to recognize the exact shapes. For this reason, we segmented these patterns using the spectral data provided by a spectral sensor.

Liquid Crystal Tunable Filter

It was necessary to take a wide range of spectral images because mural patterns were faint and the exact positions were unclear. The line spectral sensor developed for the Ozuka tumulus was only suitable for taking spectral data from a local limited region. For a wider region, the sensor has a disadvantage because the horizontal resolution changes depending on the distance of a target region. Additionally, owing to its bulky rotation mechanism, it is difficult to mount it on the automatic tripod head for omnidirectional spectral image capture. Therefore, we designed an omnidirectional spectroscope using a liquid crystal tunable filter.

(a) **(b)**

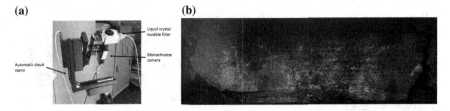

Fig. 11.7 Omnidirectional spectroscopy system. **a** Monochrome camera with a liquid crystal tunable filter mounted on an automatic tripod head. The tunable filter samples the entire visible light at 81 wavelengths according to the voltage applied. The automatic tripod head rotates the system omnidirectionally 360°. **b** An omnidirectional RGB image reconstructed from 81 omnidirectional spectral images

The spectroscopy system comprised a monochrome camera with a liquid crystal tunable filter as shown in Fig. 11.7a [5]. The tunable filter changes the transmittable wavelength according to the applied voltage. The system automatically samples the entire visible light at 81 wavelengths, by changing the voltage, resulting in a sequence of 81 images from the same viewing direction. Because the filter covers the entire field-of-view of the camera, the system provided a uniform sampling density in the horizontal and vertical direction. The resolution of one spectroscopy image was 1280×920, and each pixel had spectral information of 81 bands at intervals of 4 nm in 400–720 nm. We mounted the system with a high-color-rendering light source on an automatic tripod head.

This system automatically adjusts the exposure time at each wavelength. Because the spectral sensitivity of this system was lower at shorter wavelengths, those parts contained more noise because of insufficient light under a fixed exposure. Thus, the optimum exposure time was automatically selected according to the sensitivity characteristic of the system and the light source spectral distributions at each wavelength.

Figure 11.7b shows an omnidirectional RGB image reconstructed from 81 omnidirectional spectral images obtained by the system.

Cyber Archeology: Original Pattern Detection

The mural paintings in the Noriba tumulus were severely deteriorated. The patterns under RGB images were, therefore, difficult to recognize. We extracted the regions of the patterns from the background using the optimal wavelength of the spectral images. The spectral image had a high dimension (81 dimensions) in the wavelength direction. Additionally, normal segmentation methods were very difficult because of computational loads. However, most dimensions did not contain any meaningful information. We developed a normal cuts (an NCuts) algorithm to perform nonlinear dimensional compression and pattern-segmentation [6].

We obtained segmentation results using the NCuts algorithm from omnidirectional multispectral images from the walls of the front and back chambers [7]. From the image of the left wall of the back chamber, we found triangle patterns, as expected. See Fig. 11.8a–c. The image of the left wall of the front chamber contains the patterns as those shown in Fig. 11.8d, e. This result suggests the possibility of the existence

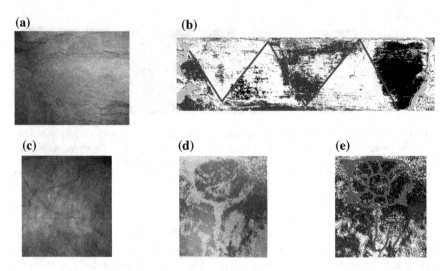

Fig. 11.8 Noriba mural painting. **a** The target area in the left wall of the back chamber (RGB) image. **b** Extracted pattern. **c** The target area in the left wall of the front chamber (RGB) image. **d** Extracted pattern. **e** Interpretation as a bi-legged ring-shaped pattern

of a bi-legged ring-shaped pattern, whose existence was not at all expected. This is a very surprising discovery because this pattern is very rare and has been confirmed only in four tumuli.

11.1.3 Sakurakyo Tumulus

The Sakurakyo tumulus is another keyhole-shaped tomb, located in Munakata-shi, built in the second half of the sixth century. This tumulus contains repeated triangular patterns with engraved lines. The lines depict square compartments and diagonals. Each triangle is painted either in red, white, or green.

Photometric Wing

We developed a photometric wing device to measure the narrow engraved lines (~ 0.5 mm). A standard range sensor could not accurately measure the narrow engraved lines. Additionally, the brightness of the reflected light needed to be sufficient for accurate measurement. Too bright of a light on one narrow area could cause damage to the target object. To protect the engraved patterns, we abandoned the high-powered Lidar sensor and developed a photometric wing device, to project weak light on a wide area, as shown in Fig. 11.9.

The device is based on an extension method of photometric stereo. As described, photometric stereo determines surface orientations from shading information provided by multiple images obtained from the same observing direction under different

Fig. 11.9 Photometric wing

illumination directions. It is easy to increase the spatial resolution by simply using a high-resolution camera.

General photometric stereo can only handle Lambertian reflection. Photometric wing, on the other hand, can handle non-Lambertian reflection [8, 9], which can be approximated as a linear combination of Lambertian reflection and the specular reflection components [10]. The ratio of both components and the observable range of the specular reflection component depends on the surface material and roughness. It is also known, however, that the observable range is relatively limited. When illuminated with six different light sources, it is certain that some of the images will have both components, whereas the remaining images will contain only the Lambertian component. By using the graph-cut method for out-layer rejection, the photometric wing selects only those images having Lambertian reflection components and determined surface orientations using those images.

Cyber Archeology: Analysis of Engraving Lines

Line engravings were normally created as drafts of mural paintings. Figure 11.10a shows a black-and-white image based on the surface orientation distribution given by the photometric wing. Figure 11.10b shows the color image of the mural painting, on which the engraving lines are superimposed from Fig. 11.10a. Apparently, these line-engraved patterns differ from the boundary of the colored areas.

The issue then becomes how to consider the cause of these gaps. Of course, because the task was performed by a human, it may have simply been accidental. However, the situation in which over-paintings occur repeated does not appear to have been accidental. In this case, it appears that paintings were made without considering the existence of the engraving lines. Namely, we must consider that two types of patterns (i.e., engraved lines and colors), coexist independently. However, we do not know the reason for this case.

(a)   **(b)**

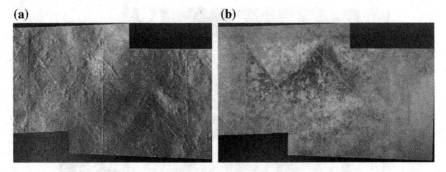

Fig. 11.10 Sakurakyo triangular patterns. **a** The black-and-white image depicts surface orientation distribution obtained by the photometric wing. This image extracts the positions of the engraving lines. **b** A color image of the pattern with superimposed engraving lines

11.2   Geometric Aspect: Cambodian Project

This section addresses two efforts to digitize the Angkor Dynasty archeological sites in Cambodia. Although physical restoration efforts have been made by many international teams under UNESCO, damage continues to progress day-by-day because of heavy rainfall and severe sunshine peculiar to the rainforest climate. For example, at the Bayon temple, at the center of the Angkor archeological site, the central tower leans farther year after year, causing fear of its inevitable collapse. Thus, we decided to digitally preserve the current shape using the recently developed e-heritage techniques (i.e., active-lighting sensors). We describe our efforts at the Bayon and the Preah Vihear temples.

11.2.1   Bayon Temple

The Bayon temple is considered the greatest masterpiece among the temples of the Angkor Dynasty. This is a stone-set temple built in the twelfth and fourteenth centuries during the dynasty of Jayavarman 7. See Fig. 11.11. The temple consists of a central tower that symbolizes Mt. Sumeru, the highest mountain rising in the center of the world in Buddhism, including 50 sub-towers with 173 existing deity faces and one outer corridor with a relief depicting people's lives and battles won against the Champa army and one inner corridor with reliefs of the theme of mythology.

Sensors for Bayon Data Acquisition

Measuring the Bayon temple, a large-scale stone structure 150 m long, 150 m wide, and 40 m high, required various types of active-lighting range sensors. Our aim was to obtain the data of the outer shape of the temple and its deity faces carved on the towers as well as the inner shape of the corridors and the reliefs along the cor-

Fig. 11.11 Bayon temple in Cambodia

ridors [11]. Therefore, we employed multiple active-lighting range sensors having different resolution and measurement ranges. The sensors included both commercially available sensors, such as Cyrax, Z+F, and Vivid sensors, and those developed by us (e.g., such as balloon, mirror, rail, and climbing sensors).

The Cyrax and Z+F sensors, the ToF-type sensors, were employed to obtain the data of the outer shape and some of the deity faces near the ground. Those sensors projected laser light to the target and measured the distance from the ToF. While the Cyrax sensor directly measured the ToF of the laser light, the Z+F sensor measured the phase difference between the outgoing and incoming laser. The Vivid sensor, a light-stripe type, was attached to our mirror reflection mechanism to extend the clearance between the sensor and the object and was employed to measure the hidden pediments.

A balloon sensor was employed to measure the deity faces at high positions. A rail sensor was used to efficiently obtain data of reliefs along the main corridors. A climbing sensor was employed to measure narrow corridor and stair walls densely existing inside of the temples.

We developed the balloon sensor for scanning the high-level deity faces [12], such as the one atop of the central tower. To do this, we hung a laser scanning unit and a camera to a floating balloon. See Fig. 11.12a. The laser scanning unit used a ToF-ranging mechanism, which projects laser light to a target and measures its distance. The laser beam scans horizontally by using a rotation mirror and vertically by using a swing mirror. This scanning unit obtains range data of 900×160 pixels per second. Owing to balloon movement during scanning, the obtained data were distorted as shown in Fig. 11.12b.

We also mounted a color camera atop the ranging unit for rectification purposes. This camera and the laser scanning unit were synchronized and spatially calibrated. Thus, the distorted range data applied color data at each pixel based on this synchro-

(a)

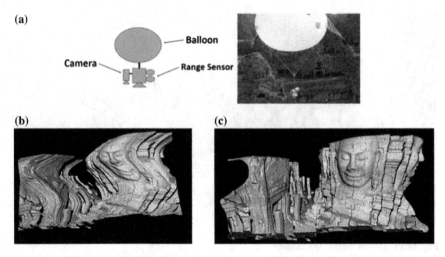

(b)   **(c)**

Fig. 11.12 Balloon sensor. **a** Sensor under the balloon. **b** Distorted range data from the sensor caused by balloon shaking. **c** Rectified range data from the algorithm

nization. We used the factorization algorithm to a sequence of color images provided by the color camera for obtaining rough 3D point-cloud data and the sensor position [13] for obtaining rough rectified data as the initial solution of the next precise rectification.

We have developed an algorithm to obtain rectified data by using both balloon-based distorted data and factorization-based rough data. The algorithm minimizes the energy function, consisting of three constraint terms: the range data constraint, the bundle adjustment term, and the smoothness constraint. This allows us to determine the camera trajectory precisely. The range data constraint minimizes the difference in 3D coordinates estimated from the factorization method and measured by the range sensing unit. The bundle adjustment term minimizes the difference between camera image points and re-projected points from the factorization 3D coordinates based on the current camera position. The smoothness constraint minimizes the difference between two adjacent camera coordinates. Figure 11.12c shows the rectified 3D point-cloud data created from the algorithm. See [12] for more details.

Bayon temple has many inner corridors where the widths are narrow enough for one person to barely get through. Usually, a range sensor has poor accuracy when measuring a distance less than 1 m. To scan such narrow areas, we developed the climbing and mirror sensors [14]. The climbing sensor was used to scan narrow hidden stairs, and the mirror sensor was used to scan pediments intentionally hidden between two high walls.

Thus, we obtained more than 10,000 range images with 0.25 TB of range data.

Software Pipeline to Process Bayon Temple Data

A standard data acquisition process requires three steps: partial data acquisitions, alignments, and merging into a mesh model of uniform spatial resolution. During data acquisition, a single scan only provides partial data of the object because of occlusions and/or limitations of the field of view. Multiple times of partial scans are, therefore, conducted to produce a set of partial data that covers the entire surface of an object. Then, the spatial relations among those data are determined by using an alignment algorithm, such as the Iterative Closest Point (ICP) algorithm [15]. Finally, those partial data are connected based on the obtained spatial relations into a geometric model of the object by using a merging algorithm, such as the marching cube algorithm [16].

The scale of the temple prevented from the use of existing algorithm pipelines. Each ICP process provides a small error and, by repeating ICP 10,000 times, those small errors accumulate gradually. Thus, when we finish all the alignment processes, a large gap occurs between the first data and the last data. To avoid this accumulation error, standard simultaneous alignment algorithms have been developed to load all point-cloud data into the computer and consider all possible combinations among the data for alignment. However, such standard simultaneous algorithms cause memory overflows and combinatory explosions when handling the Bayon dataset.

We developed a two-step alignment algorithm. The first step is a graphics processing unit (GPU)-based pairwise high-speed algorithm that roughly aligned the data obtained at the site [11, 17]. In particular, the GPU painting function was used for a time-consuming point correspondence search. One point-cloud datum is converted to a mesh datum and each triangular mesh is colorized with a unique color by using the GPU painting function. This colorized mesh datum is projected as an index image to the image plane perpendicular to the scanning direction of the second point-cloud data. Then, the second point-cloud datum is projected on this index image, and each point is colorized based on the first mesh color of the image plane by using the GPU painting function. From the color information, we can obtain the correspondence between the first and second point-cloud data. Based on the point correspondence, we can then determine the relative relationship between the two point-cloud data. These correspondences may contain a small percentage of miscorrespondence caused by occlusion. However, at this first step, we ignore them because most correspondences provide relatively accurate alignment results.

The second step of the alignment algorithm runs in parallel on a PC cluster [18]. Theoretically, from a given set of point-cloud data, simultaneous alignment algorithms must consider all possible combinations of overlapping adjacent relations among all point-cloud data. Fortunately, the first step already established rough alignment and provided adjacent relations. The second step makes it possible to reduce computation time and the amount of memory used by removing unnecessary relations and clustering them into dependency groups. Because the alignment computation between two point-cloud data can be performed independently, each dependency group is assigned to each PC node. Because the computation time is proportional to the number of points assigned to each node, by forming those clusters so that the number of points computed is equal on each node, the load at each node is efficiently distributed.

After aligning all point-cloud data, it then becomes necessary to merge them into a complete mesh representation with uniform spatial resolution. If we simply connect

Fig. 11.13 Digital Bayon (Complete mesh model)

all mesh dates given by point-cloud data, the final result will not be uniform because of the differences in the resolution of each point-cloud datum. Additionally, the final result may contain holes. To remedy these issues, we developed a parallel merging algorithm to move all point-cloud data into one space [19]. Then, the algorithm divided this space into small voxels, and obtained the probability of the existence of a surface in each. By connecting voxels having a high probability of surface existence using the marching cube algorithm [16], we obtained the final mesh representation of the Bayon temple. This merging operation was also conducted using a PC cluster in parallel. Figure 11.13 shows the merged result of the Bayon temple. The temple 150 m long, 150 m wide, and 40 m high was represented in a spatial resolution of 2 cm in each direction. The original data had more accuracy, but, owing to the limitation of the memory space of our PC cluster, we ended up this spatial resolution.

The obtained Bayon 3D mesh model can be used in various applications. One included the creation of video contents. One Japanese company created a virtual-reality system to allow users to virtually visit the Bayon temple by using the mesh model. A detailed mesh model with 2 cm resolution was not necessary for this application. One application that requires such a highly accurate 3D mesh model is archeological analysis. In the next section, we introduce this application.

Cyber Archeology: Deity Face Analysis

We conducted archeological analysis on the Bayon deity faces using the 3D mesh data. About 50 sub-towers around the central tower have four engraved deity faces each. Currently, 173 existing faces are digitized by using the balloon sensors, and those data are included in the Bayon 3D mesh model. According to art experts, the faces are roughly classified into three groups: Deva, Asura, and Devata groups. However, this classification is based on subjective judgment. Moreover, some faces among the 173 existing faces are difficult to classify from their appearances because of dirt from vegetation and/or partial defects.

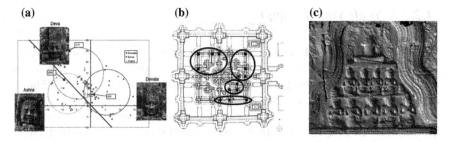

Fig. 11.14 Bayon cyber-archeological analysis. **a** Supervised analysis of 173 faces. **b** Four carving teams. **c** Synthesized pediment and converted statue image

We ran supervised classification on all 173 faces. We normalized the data based on facial landmarks so that all had roughly the same size and depth. Then, we ran supervised analyses of discrimination functions from experts' classification based on faces without ambiguity. We then obtained the face distribution in the low-dimensional space using the discrimination functions shown in Fig. 11.14a. A point corresponds to one of the 173 faces. The vertical line indicates the discrimination function between a female god (Devata type) and a male god (Deva and Asura type). The diagonal line indicates the discrimination function between Deva and Asura type.

We also ran unsupervised classification on them. From this analysis, we can identify four similarity groups and their existing proximity positions, as shown in Fig. 11.14b. There is a legend that multiple teams carved the deity faces in parallel. From our unsupervised analysis and this legend, we can infer that at least four carving teams worked in a parallel manner. The first team carved the four sub-towers at the northwest corner. The second team did the four sub-towers at the northeast corner. The third team did three sub-towers in the central area. The fourth team did the three sub-towers at the southeast corner.

Bayon temple has many pediments. However, via renovations, ancient engineers may have intentionally hidden them behind walls because of religious changes. Most cannot be observed, even today. Narrow clearances of 30 cm are the distances between the pediments and the other side of the wall, located at a height of about 5 m. There were no pictures prior to our digitization. By placing a mirror diagonally, with respect to the pediment and vivid range sensor, we obtained a small piece of point-cloud data of the pediments at every 10 cm. Then, we merged them into a complete dataset and synthesized pictures of all existing pediments.

Figure 11.14c shows one such synthesized picture. In the upper central area, one can observe a vertical bar. According to an art expert, the original Buddha statue that once existed there was scraped off and re-carved into a Linga statue, the symbol of Hindu Shiva god. For this reason, we can surmise that this temple may have been converted from a Buddhist temple to a Hindu temple.

11.2.2 Preah Vihear Temple

The Preah Vihear is a Hindu temple, located at the border of Cambodia and Thailand. The construction began at the end of the ninth century by the Angkor Dynasty and is believed to have been completed by the eleventh century. It was built on a cliff in the Dngrłk mountains with a magnificent view of Cambodian to the south. The approach road extends from the Thailand side from the north to the south, going through the five gopuras (Gopura I to Gopura V), and finally through the main courtyard at Gopura I.

We obtained a 3D point-cloud data of this temple to help preserve the structure virtually, owing to the possibility of physical destruction via international conflicts, and to capture the uniqueness of the north–south main axial direction of the temple [20].

Topographically, the old approach road reaches to the temple from Thailand. However, from negotiation between the Thai and French, who occupied Cambodia at the time, it was agreed that the temple belonged to Cambodia. As a result, the temple cannot be reached from the original approach road. Several military clashes between Thailand and Cambodia have occurred near the temple since Cambodia's independence. Fortunately, both sides seemed to actively avoid damaging to the site. Nevertheless, such preservation cannot be guaranteed. As a matter of fact, there are bullet holes in the wall of the temple from the previous conflict. We, therefore, considered it is worthwhile to obtain 3D point-cloud data of Preah Vihear temple before it is destroyed.

The temple has a unique position among the Khmer temples built during the Angkor Dynasty. Although most Khmer temples, including Angkor Wat and Bayon, are based on the east–west main axis, this temple is based on a north–south axis. We conjecture that this was intentional. There is an ancient legend that Preah Vihear temple overlooks important Khmer temples, as the guardian temple of the Angkor dynasty. We used 3D point-cloud data to search for cues about this legend.

The point-cloud data of Preah Vihear temple was obtained by the data acquisition team from the University of Tokyo (UTokyo) from 2012 through 2015 in cooperation with Japan and Apsara safeguarding Angkor (JASA) Team and the Preah Vihear Authority. We used active-lighting laser range sensors developed by UTokyo and commercially available sensors with UTokyo-developed software. This section briefly describes these sensors and software tools and presents a visualization of the obtained point-cloud data.

3D Active-Lighting Laser Range Sensors

We used commercially available Imager5010c (manufactured by Z+F) and C10 (manufactured by Leica Geosystems). The maximum measurement distances of Imager5010c and C10 are 187 and 300 m, respectively. These sensors can obtain omnidirectional point-cloud data at one scan by emitting laser light in the vertical directions while horizontally rotating the sensor-head.

Because Preah Vihear is built on a ridgeline of the Dngrłk mountains, it is important to archive the terrain around the buildings to obtain the complete 3D point-cloud

Fig. 11.15 Omikoshi-type handcart sensing system developed for the Preah Vihear temple and the obtained data. The Omikoshi-sensor system consisted of a laser profiler and omnidirectional camera

data representing the entire temple. Unfortunately, the aforementioned sensors were not suitable for measuring wide areas, because they must be fixed on the ground with a tripod. Furthermore, it requires too much time for scanning (several minutes). A commercially available mobile sensor, such as Velodyne, has a much coarser resolution that is not suitable for our digital archiving purpose.

We, therefore, developed a handcart 3D-measurement system that can measure wide terrains from the ridge line. Figure 11.15 shows the measurement handcart with a laser profiler and an omnidirectional camera mounted. We used a laser profiler for shape measurement, which measured 3D information with respect to the sensor position in a plane perpendicular to the moving direction. Movement of the platform provided the next scanning area of the profiler. We chose a human-driven handcart for careful and safe manuring over uneven terrains surrounded by steep cliffs around the temple buildings.

Platform motion was estimated by using the images taken by the omnidirectional camera. The relative sensing positions in the moving direction are unknown. For estimating the motion of the sensor while scanning, feature points in the images were tracked through sequential images. Then, applying the factorization method to those feature-point images, the motion and depths of the feature points were estimated [13]. The depths obtained by the laser scanner were used to estimate the absolute scale of the motion and to refine the accuracy by assuming that the cart movement was along a straight line for a short interval (a couple of seconds) [21].

We obtained a complete 3D point-cloud data of Preah Vihear temple, as shown in Fig. 11.16. Note that owing to the effective scanning by the handcart scanning system, not only five gopuras (Khmer-style temple gates) but also the surround-

Fig. 11.16 3D point-cloud data of Preah Vihear temple. Note that owing to the effective scanning of the handcart scanning system, not only five gopuras but also surrounding terrains are included in this point-cloud data. Due to the book layout, the left direction indicates the south direction (Cambodian side) and the right direction indicates the north direction (Thailand side)

(a)   **(b)**   **(c)**

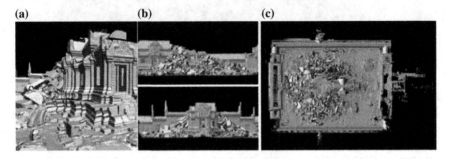

Fig. 11.17 3D mesh model of Gopura I. **a** Buildings of Gopura I. **b** Central sanctuary of Gopura I. **c** Orthogonal image

ing terrains, were included in this point-cloud data. Owing to the book layout, the left direction indicates the south direction (Cambodian side) and the right direction indicates the north direction (Thailand side). Ancients pilgrimages approached this temple from the stairs at the right side, which went through the approach road from Gopura I through Gopura V on the left side. They overlooked the magnificent view of Cambodian plain further south from the cliff edge of the Dngrĭk mountains.

Figure 11.17 shows the measurement results of Gopura I. Because this gopura has a main building surrounded by corridors. We placed the sensors inside of the building and on the roofs of the central sanctuary and the corridors. Moreover, for future virtual reconstruction purposes of the central sanctuary, we measured fallen stones as shown in Fig. 11.17b. As shown in Fig. 11.17c, we could easily generate orthogonal images of the buildings from the 3D models. This orthogonal image was only possible from the 3D mesh model, because, under traditional measurement methods, it is difficult to connect the parts of the same sea level because of elevation changes.

Figures 11.18 and 11.19 show the 3D point-cloud data of the Gopuras II and III, respectively.

Figure 11.20 shows the 3D mesh model viewed from the east, the west, the north, and the south as well as the overview directions.

Fig. 11.18 3D point-cloud data of Gopura II

Fig. 11.19 3D point-cloud data of Gopura III

Figure 11.21 shows the 3D mesh model of Gopura V, which has a beautiful gate, and the platform of the building remains. Unfortunately, most of the building has collapsed. We measured the gate, fallen stones, and remaining pillars from the ground, and from the top of the pillars.

Cyber Archeology: Spatial Arrangement of Khmer Temples

This subsection examines the spatial distribution of the Khmer temples with respect to Preah Vihear by using obtained 3D point-cloud data. First, we determined the direction the temple was facing. For this purpose, we obtained the transformation from the model coordinates on which current 3D point-cloud data is represented into the Universe Mercator (UTM) coordinates. Then, we examined the spatial arrangement of Khmer temples in UTM coordinates.

From Model Coordinates to UTM Coordinates

Because the 3D point-cloud data are represented by an arbitrary model coordinate system, the data must be aligned into a world coordinate system by using Global

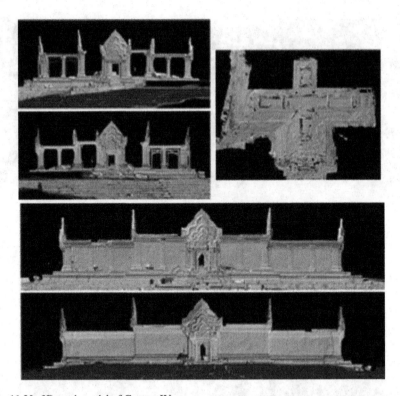

Fig. 11.20   3D mesh model of Gopura IV

Fig. 11.21   3D mesh model of Gopura V

Positioning System (GPS) data at the reference points. To analyze the spatial relationship among other Khmer temples, it was necessary to convert these coordinates into UTM, because their spatial positions were represented in UTM. For this purpose, we obtained an affine transformation, which was used to convert the model coordinates into UTM by using the reference points, of which, both the model and the UTM coordinates were known.

At Preah Vihear, 7 Preah Vihear GPS (PVG) reference points and 17 Preah Vihear Total station (PVT) reference points were installed and measured by the JASA team. PVG reference points were installed in well-viewed areas without any obstacles, and their UTM coordinates were obtained directly using the GPS signals over reasonably long observation periods.

PVT reference points were obtained by surveying multiple PVG reference points using a total station. Because all PVGs were located in well-viewed locations, they were all far away from the temple. PVT reference points were introduced to obtain UTM coordinates at key feature positions near or inside the temple. We obtained their coordinates by manually examining the 3D point-cloud data with a MeshLab user interface.

Let $\mathbf{x} \in \mathbb{R}^3$ be the model coordinates of a reference point and $\mathbf{y} \in \mathbb{R}^3$ be the UTM coordinates of the same point. We can relate these two coordinates using an affine transformation:

$$\mathbf{y} = \mathbf{Ax} + \mathbf{b}, \tag{11.3}$$

where $\mathbf{A} \in \mathbb{R}^{3 \times 3}$ and $\mathbf{b} \in \mathbb{R}^3$ are the linear transformation component and the translation component of the affine transform, respectively.

To determine $\mathbf{A}$ and $\mathbf{b}$, we need multiple pairs of model and GPS coordinates. When m data points are available, $\mathbf{A}$ and $\mathbf{b}$ in Eq. (11.3) are determined as

$$(\mathbf{A}, \mathbf{b}) = \underset{(\mathbf{A},\mathbf{b})}{\operatorname{argmin}} \sum_{i=1}^{m} \|\mathbf{y}_i - \mathbf{Ax}_i - \mathbf{b}\|^2. \tag{11.4}$$

The obtained parameters of the affine transform are as follows:

$$\mathbf{A} = \begin{pmatrix} -5.7546 \times 10^{-1} & -8.1775 \times 10^{-1} & 3.6220 \times 10^{-1} \\ 8.2985 \times 10^{-1} & -5.4907 \times 10^{-1} & -9.3693 \times 10^{-3} \\ 5.7153 \times 10^{-3} & 1.3311 \times 10^{-2} & 9.9203 \times 10^{-1} \end{pmatrix},$$

$$\mathbf{b} = \begin{pmatrix} 4.6553 \times 10^5 \\ 1.5909 \times 10^6 \\ 6.5291 \times 10^2 \end{pmatrix}.$$

Features to be Used for Determining the Major Direction of the Temple

One issue related to determining the direction of the archeological site is that of determining which feature will be employed for the purpose. One popular method in modern construction is to directly use reference points located on such sites for determining site directions. This method first sets reference points somewhere in the site and then measures the UTM coordinates of these reference points. This was the case for PVGs and PVTs in the previous section. Then, based on these UTM coordinates, we can determine the direction of the site. However, this direct method is susceptible to errors, depending on where reference points are placed at the site. Modern construction sites, consisting of fresh straight lines, provide ample candidate points. However, this method has difficulty handling ancient leaning stone building with swaying without any clear boundary edges.

The second method determines the contour lines of the site. This method is also often used in modern construction sites having clear boundaries. However, Preah Vihear temple does not have clear boundary lines, and some parts are half-buried or well-deteriorated. Even if the point-cloud data are available, it is difficult to extract clear boundaries.

The third possibility determines the principal axis of the point-cloud data of all buildings. However, the axis direction obtained varies, depending on how we segment the point-cloud data into buildings.

From these considerations, we decided to extract the direction of the walls from the point-cloud data. Fortunately, the Preah Vihear temple has a number of long walls either along the axis or perpendicular to it. Because the number of points in the data was sufficiently large, the direction of the walls was determined stably, without depending on how many points we extract from the point-cloud data for this calculation when the rough wall areas are manually extracted. The next section will explain this method in detail.

Axis Direction and its Interpretations

We conducted the following two steps for axis determination. In the first step, the walls to be used for determining the axis direction were manually segmented, and their normal directions were obtained by applying Principal Component Analysis (PCA) to the point-cloud data of the segmented regions. We use MeshLab [22] for manual segmentation of the walls. Since the number of the points in the extracted regions are sufficiently large, the direction of the walls can be determined stably without much depending on how many points we extract from the point-cloud data for this calculation once rough wall areas are manually extracted.

The second step obtained the axis direction of the entire temple in model coordinates from the weighted sum of the normals obtained from the point-cloud data of the walls. As a result, the direction of the central axis was determined to be tilted 3.9212° to the west with an error range of 1.4°, compared to the true south. We considered the tilt to be within the range of technical errors at that time and, thus, the Preah Vihear temple was built along the north–south line.

Revisiting the Khmer legend that stated Preah Vihear temple was built facing Angkor Wat temple, as the guardian temple of the Angkor Dynasty, the above result clearly rejects the legend, because the line connecting Angkor Wat to Preah Vihear has an angle of approximately 40° from the north–south line.

Table 11.1 List of temples that were considered in this paper. All temples are UNESCO's World Heritage sites

ID	Temple name	Construction period	Dynasty	Note
1	Sambor Prei Kuk	600-	Isanavarman I	The oldest site has an octagonal hall has a north–south axis
2	Preah Vihear	890–1150	Suryavarman I Suryavarmann II	
3	Vat Phou	900-	Yasovarman I	Located in the southern Laos
4	Koh Ker	921–944	Jayavarman IV	Pyramid style
5	Banteay Srei	967-	Rajendravarman II Jayavarman V	Oriental Mona Lisa
6	Bayon	1100-	Jayavarman VII	Referred to as the center of the universe
7	Beng Mealea	1100-	Suryavarman II	Referred to as East Angkor Wat
8	Angkor Wat	1113–1145	Suryavarman VII	The symbol of Cambodia
9	Ta Prohm	1186	Jarvarman VII	Buddhism to Hinduism entangled by big trees
10	Banteay Chhmar	1100–1200	Jarvarman VII	Near the border of Thailand has a thousand-armed Avalokiteshvara

Next, we consider the spatial arrangement of Khmer temples based on this direction. For this purpose, we selected representative UNESCO-registered World Heritage Khmer temples from nearby areas, as shown in Table 11.1.

The first conjecture is that the temples were placed in 24 directions. The Chinese 24 solar term reflects the Chinese calendar, which divides 1 year into 24 seasons according to the positions of the sun. This calendar has been used since the Chinese Warring States Period (BC403–BC221). All important framing events were conducted according to this 24 solar term. The interpretation of the calendar is considered the symbol of the emperor's authority. Figure 11.22a shows the results of the 24 lines drawn radially every 15° from the Preah Vihear. Markers indicate the locations of the Khmer temples in Table 11.1. Some important Khmer temples are located on these lines but not all.

Another conjecture draws nine lines around the temple. The line connecting Preah Vihear and Angkor Wat, the most important temple of the Angkor Dynasty, lies 40° with respect to Preah Vihear main north–south axis. Dividing 360 by 40 is, in fact, nine. Moreover, number nine, is an important number in Hinduism and Buddhism. In Buddhism, the number represents god and universe. Although present temples

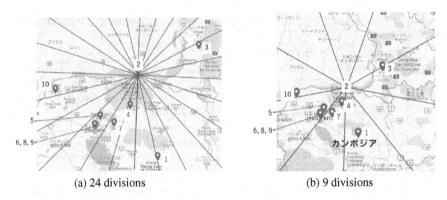

(a) 24 divisions   (b) 9 divisions

Fig. 11.22 Hypothetical interpretation of the spatial arrangement of the archeological sites

are the faith objects for Buddhism, they were Hindu temples at the time of their constructions. Among the Hindu gods (e.g., Vishnu, Shiva, and Brahma) the central Sun God, Vishnu is the most important. He has 10 different human forms and his ninth form is the Buddha. Thus, the nine is also considered as important to Hinduism. Figure 11.22b shows these nine lines. Again, some important Khmer temples are located on these lines but not all.

From this analysis, we rejected the Khmer legend that the Preah Vihear temple, as the guardian temple of Angkor Dynasty, was built to face Angkor Wat or Bayon. Although many important temples are located on 9 or 24 division lines, we cannot conclude that all temples fell into either system. Further analysis is necessary.

References

1. Ikeuchi K, Miyazaki D (2008) Digitally archiving cultural objects. Springer Science & Business Media
2. Ikeuchi K (2015) All of Kyushu decorative tumuli. Tokyo Syoseki. (in Japanese)
3. Masuda T, Yamada Y, Kuchitsu N, Ikeuchi K (2008) Illumination simulation for archaeological investigation. In: Ikeuchi K, Miyazaki D (eds) Digitally archiving cultural objects, chap 10. Springer, pp 419–439
4. Ikari A, Masuda T, Mihashi T, Matsudo K, Kuchitsu N, Ikeuchi K (2005) High quality color restoration using spectral power distribution for 3d textured model. In: 11th international conference on virtual systems and multimedia
5. Morimoto T (2009) Reflectance analysis of layered surfaces using a multispectral image. PhD thesis, Graduate School of Information Science and Technology, the University of Tokyo
6. Morimoto T, Mihashi T, Ikeuchi K (2006) Color restoration based on spectral information using normalized cut. ITE Trans Media Technol Appl 62(9):1–6 (in Japanese)
7. Morimoto T, Kuchitsu N, Ikeuchi K (2011) Pattern detection of mural paintings using panoramic multispectral camera. In: 2011 Annual meeting, Japan society for scientific studies on cultural properties. (in Japanese)
8. Miyazaki D, Hara K, Ikeuchi K (2010) Median photometric stereo as applied to the segonko tumulus and museum objects. Int J Comput Vis (IJCV) 86(2–3):229

9. Miyazaki D, Ikeuchi K (2010) Photometric stereo using graph cut and m-estimation for a virtual tumulus in the presence of highlights and shadows. In: Proceedings of the IEEE conference on computer vision and pattern recognition (CVPR) workshops. IEEE, pp 70–77

10. Nayar SK, Ikeuchi K, Kanade T (1991) Surface reflection: physical and geometrical perspectives. IEEE Trans Pattern Anal Mach Intell (PAMI) 7:611–634

11. Ikeuchi K, Oishi T, Takamatsu J, Sagawa R, Nakazawa A, Kurazume R, Nishino K, Kamakura M, Okamoto Y (2007) The great buddha project: digitally archiving, restoring, and analyzing cultural heritage objects. Int J Comput Vis (IJCV) 75(1):189–208

12. Banno A, Masuda T, Oishi T, Ikeuchi K (2008) Flying laser range sensor for large-scale site-modeling and its applications in bayon digital archival project. Int J Comput Vis (IJCV) 78(2–3):207–222

13. Tomasi C, Kanade T (1992) Shape and motion from image streams under orthography: a factorization method. Int J Comput Vis (IJCV) 9(2):137–154

14. Matsui K, Ono S, Ikeuchi K (2005) The climbing sensor: 3-d modeling of a narrow and vertically stalky space by using spatio-temporal range image. In: Proceedings of the IEEE/RSJ international conference on intelligent robots and systems (IROS). IEEE, pp 3997–4002

15. Besl PJ, McKay ND (1992) Method for registration of 3-d shapes. In: Sensor fusion IV: control paradigms and data structures, vol 1611. International Society for Optics and Photonics, pp 586–606

16. Lorensen WE, Cline HE (1987) Marching cubes: a high resolution 3d surface construction algorithm. In: Proceedings of SIGGRAPH, vol 21. ACM, pp 163–169

17. Oishi T, Kurazume R, Nakazawa A, Ikeuchi K (2005) Fast simultaneous alignment of multiple range images using index images. In: Fifth international conference on 3-d digital imaging and modeling (3DIM'05). IEEE, pp 476–483

18. Oishi T, Nakazawa A, Sagawa R, Kurazume R (2003) Parallel alignment of a large number of range images. In: Proceedings of the fourth international conference on 3-d digital imaging and modeling, 2003. 3DIM 2003. IEEE, pp 195–202

19. Sagawa R, Nishino K, Ikeuchi K (2001) Robust and adaptive integration of multiple range images with photometric attributes. In: Proceedings of the IEEE conference on computer vision and pattern recognition (CVPR), vol 2. IEEE, pp II–II

20. Kamakura M, Ikuta H, Zheng B, Sato Y, Kagesawa M, Oishi T, Sezaki K, Nakagawa T, Ikeuchi K (2019) Preah vihear project: obtaining 3d point-cloud data and its application to spatial distribution analysis of khmer temples. In: 3rd ACM SIGSPATIAL international workshop on geospatial humanities, Chicago, IL

21. Zheng B, Oishi T, Ikeuchi K (2015) Rail sensor: a mobile lidar system for 3d archiving the bas-reliefs in angkor wat. IPSJ Trans Comput Vis Appl 7:59–63

22. Cignoni P, Callieri M, Corsini M, Dellepiane M, Ganovelli F, Ranzuglia G (2008) Meshlab: an open-source mesh processing tool. Eurograph Italian Chap Conf 2008:129–136

Chapter 12
Robot Vision, Autonomous Vehicles, and Human Robot Interaction

Sensors are indispensable to robots. Active-lighting sensors, for example, work in a robust manner, even under severe conditions. They are utilized heavily for various robot tasks, such as object localization/recognition, navigation, and manipulation. In this chapter, we review two robotics applications: simultaneous localization and mapping (SLAM) for navigation tasks and learning-from-observation (LfO) for manipulation tasks.

12.1 Simultaneous Localization and Mapping (SLAM)

As humans move around making decisions based on visual cues gathered from their eyes, robots are required to move around based on visual cues gathered from their visual sensors. This functionality is called *navigation*. One representative example of machine-supported navigation is autonomous driving. For such navigation tasks, 3D maps are indispensable.

This section focuses on the application of active-lighting sensors to making 3D maps using simultaneous localization and mapping (SLAM). SLAM incrementally builds a map by repeatedly estimating the robot's position while moving around. SLAM can be implemented using passive sensors or active-lighting sensors. Generally, however, it is more robust to use an active-lighting sensor. Here, we concentrate on SLAM 3D map-making using point-cloud data [1] obtained from an active-lighting sensor: LiDAR.

Generally speaking, SLAM consists of the following three processes:

- **Observation**

    - Intake new point-cloud data.

© Springer Nature Switzerland AG 2020

K. Ikeuchi et al., *Active Lighting and Its Application for Computer Vision*,
Advances in Computer Vision and Pattern Recognition,
https://doi.org/10.1007/978-3-030-56577-0_12

(a)   **(b)**

Fig. 12.1 Kanaya tunnel and measurement platform

- **Data association**

  - Compare the previous point-cloud data with the newly acquired data.
  - Extract a common set of features.
  - determine the relationship between the previous acquisition position and the current position based on the correspondence of the common set of features.
  - Merge the current point-cloud data into the previous data.

- **Loop closure**

  - After repeating the data association steps, if loop closure is detected, then update all previous positions so that the gap between the initial and final positions are minimized. With this updating process, all point-cloud data are remerged.

12.1.1   Stop-and-Go Method

We obtained the 3D point-cloud data of the 4.6-km-long Kanaya tunnel, as shown in Fig. 12.1a, using a version of SLAM, which is a stop-and-go method [2]. GPS does not penetrate tunnels. Therefore, as a means of localization for autonomous driving, the role of the 3D map inside tunnels is extremely important. However, the standard way to produce a 3D map utilizes a mobile measuring platform that utilizes GPS signals for localization. Thus, it is difficult to apply this method to tunnel-map production. To remedy this chicken-and-egg problem, we used SLAM to obtain 3D maps.

SLAM begins with data observation. In the described research, we obtained 3D point-cloud data using a Z+F imager 5030 LiDAR sensor, an active-lighting sensor, mounted on a measurement vehicle, as shown in Fig. 12.1b. This sensor can measure one omnidirectional range datum at each observation in the stop-and-go method.

SLAM repeats observation steps and simultaneously performs data association of the newly acquired data with previously acquired data. In our case, at each step, the vehicle advanced about 20 m and acquired a new omnidirectional point-cloud data. The data association process determined the relationship between two sets of data. Generally, the relationship can be determined from the point correspondence using an algorithm, such as the Iterative closest point (ICP) algorithm [3].

Data association was performed via the correspondence of 3D edges, not the correspondence of simple 3D points, as is done with ICP. Owing to the peculiarity of tunnel shapes, i.e., nearly cylindrical shapes, almost all point features resemble one another. Thus, ICP can be easily trapped in to local minima. Therefore, we extracted 3D edges from the point-cloud data. We mapped each 3D datum to a spherical surface, of which the center was the data-acquisition point. We expanded the projected line thickness over the spherical surface, 20° across edges, and remapped those regions to the original point cloud, resulting in regions at both sides of the 3D edges. See Fig. 12.2a. We conducted the ICP algorithm using only data points around those extracted 3D edges [2].

A loop closure process was necessary to complete the data association. Generally, when SLAM finishes taking all data and returns to the same place, the first and last positions deviate because of the accumulation of errors during the iterative process. The loop closure fine-tunes overall positions to eliminate this deviation. In our case, measuring the tunnel one way did not return the platform to its original position. However, GPS data exists at the entrance and the exit, providing start/stop positioning information. Each data association process in the tunnel measurement procedure was fine-tuned based on this global location fine-tuning.

For the loop-closure process, we used high-accuracy GPS data at the entrance and exit of the tunnel. We placed the disk-shaped GPS antenna at 14 places at both the entrance and exit and recorded the positional information. In parallel, we scanned the scene geometry, including the antenna, and aligned them locally. To find the position of GPS antenna, we aligned a 3D model of the antenna created by using high-precision range data with the scene-geometry range-data. We fixed both ends of the model to absolute positions. Then, keeping this constraint, we applied the simultaneous alignment algorithm [4].

Figure 12.2b shows some of the resulting 3D scenes. To evaluate the results, we compared the obtained model with the planning diagram, as shown in Fig. 12.2c. The vertical axis depicts the difference, and the horizontal difference reflects the mileage through the tunnel. The difference has the maximum value in its central part. This is because of the error accumulated in the central part although it was restrained in the entrance and the exit. Figure 12.2d shows the curvature distribution of mileage along the tunnel. The second largest difference exists around $x = -38,000$– $x = -377,000$. This area has the sharpest curvature as shown in Fig. 12.2d.

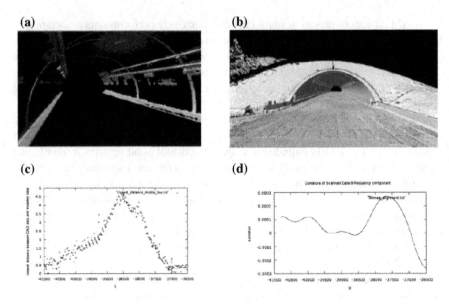

Fig. 12.2 3D model of Kanaya tunnel. **a** 3D edges extracted. **b** 3D model. **c** Difference between 3D model obtained with a blueprint. **d** Curvature distribution along the mileage

12.1.2 Continuous Horizontal One-Line Scanning Method

The method of the previous section took the stop-and-go approach for data association. Therefore, to obtain data in a busy town, it becomes necessary to stop traffic in the vicinity of the platform. This section explains a method in which data can be taken continuously by mounting a line-scanning active-lighting range sensor [5]. This makes it possible to finish data collection at higher speeds, owing to the short scanning time. We mounted two line-scanning range sensors, as shown in Fig. 12.3a. The vertical line-scanning sensor provides 3D point-cloud data on one plane while the horizontal line-scanning sensor provides features for data association.

By collecting the results of the horizontal scan, we obtained a spatiotemporal range image, as shown in Fig. 12.3. Note that the figure looks similar to a 2D epipolar image (EPI) [6]. However, it is, in fact, a collection of range images.

The features of a spatiotemporal range image represent both spatial characteristics of an object and temporal continuity derived by scanning with continuous movement of the sensor. Because the horizontal scanning sensor obtains the same object shape multiple times at multiple positions at very short intervals, the data association is relatively easy by searching a similar profile in vicinity areas.

By associating a similar shape in the spatioimage, we can obtain spatiotemporal edges. The slope of an edge represents the velocity of the sensor as follows:

$$m = \frac{\delta y}{\delta x} = \frac{-kF_0\delta t}{-\delta X} = \frac{fF_0}{V},$$

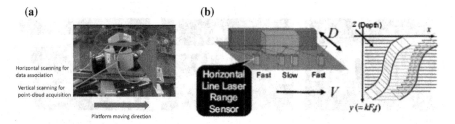

Fig. 12.3 Horizontal and vertical line scanning. **a** Data-acquisition platform. **b** The result of horizontal line-scanning

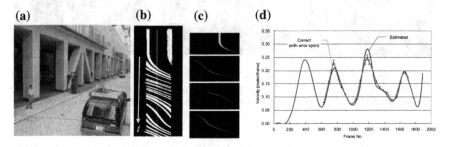

Fig. 12.4 Segmentation result. **a** Data-acquisition scene using the platform vehicle in front of the main building of Institute of Industrial Science, the University of Tokyo. **b** Data obtained with the platform. The horizontal axis represents the pixel coordinates in the horizontal scan sensor. The vertical axis represents the temporal axis. The time flows downward. **c** The segmentation result. Each cluster was extracted. **d** Estimated velocity

where V, δX, and F_0 are the velocity, moving distance, scanning ratio of the sensor, respectively. k is the interval between each frame for stacking them along the temporal axis. The slope at the edge is related to the moving velocity of the sensor, V, in the previous equation. Because each point in the spatiotemporal range image comprises cluster planes, we separate these planes. From these, we can estimate the velocity of the sensor.

We created an outdoor experiment to confirm that our algorithm worked well. We took a range image of a building at Institute of Industrial Science, UTokyo, as shown in Fig. 12.4a. We assumed the sensor platform moved parallel to the wall of the building in a straight line. The data-acquisition platform moves at the varying velocity 10–30 km/h.

Figure 12.4 depicts the results. Figure 12.4b shows the front view of the obtained spatiotemporal range image. We clustered the segmentation result into each cluster as shown in Fig. 12.4c. This segmentation associated the data association using the standard SLAM procedure. Owing to the dense sampling of horizontal scanning, this process was relatively easy.

We fitted a 6D polynomial function to each edge from the clusters to extract the velocity of the moving platform. At each period along the vertical axis, we estimated the velocity of the moving vehicle from the slope of the curve. Figure 12.4c shows

(a)

(b)

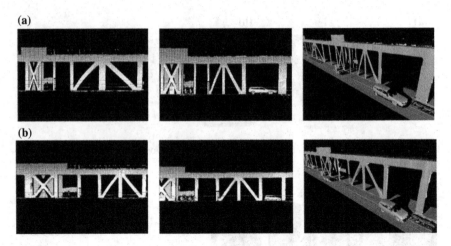

Fig. 12.5 Obtained 3D data. **a** The vertical scanning result before the data association. The vertical scanning data are aligned without considering the variance of the platform velocity. **b** The vertical scanning result based on the platform velocity given by the data association

the variation of the speed of the platform. Estimation results approximately agreed with the actual data and indicated the effectiveness of the method.

After we obtained the velocity of the platform from the horizontal scan, we rectified the result from the vertical scan so that it reflected the actual 3D world. Figure 12.5a depicts the original data from the vertical scanning sensor, and Fig. 12.5b shows the rectified result.

12.2   Learning-from-Observation

Most parts of the human brain are related to the functionalities of the hands. The main one is manipulation, which is one of the most important issues in robotics. Usually, human programmers are tasked to manually program robot programs for object manipulation. In the case of humans, it is known that manipulation tasks are obtained through learning and practices.

This section explains the learning-from-observation technique, a robot capability to learn robot tasks by observing a human performing the same tasks. Unfortunately, owing to the kinematic and dynamic differences between humans and robots, the direct mimicking method does not work well. The system first extracts the abstract meanings of human actions. Then, it maps these meanings to robotic motions. See Fig. 12.6. We explain how to represent such meanings and how to map such meanings to robot motions in various domains.

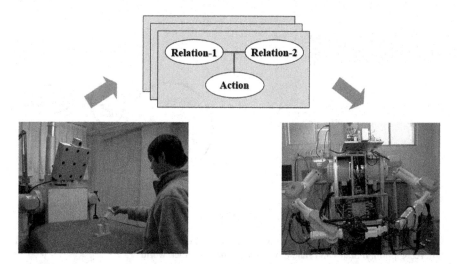

Fig. 12.6 Learning-from-observation [7]. A human demonstrates tasks in front of a robot. The robot extracts the abstract meanings of the tasks. Then, the robot maps such abstract meanings to its own actions. This indirect mapping overcomes the difficulties due to the physical difference between the human and robot bodies

12.2.1 Tasks and Skills

In the learning-from-observation paradigm, the central issue is representing the abstract meanings of actions. For this purpose, we propose the use of the task-and-skill framework. The framework separates what to do (*tasks*) from how to do it (*skills*) [8]. First, the paradigm aims to recognize the kind of a task the demonstrator is conducting such as putting an object on the table or inserting an object into a hole. For this, the system prepares a collection of tasks, representing state transitions in a certain domain. States are defined to not overlap and to cover over the domain completely.

For each task model, we prepare skill parameters to specify how to complete the task. These parameters are obtained after one input image segment is considered as a task. Examples of skill parameters include the trajectory to be executed between two states or the grasping position of an object.

As an example, let us consider the assembly of a pair of cubes with the aim of achieving specific contact states, as shown in Fig. 12.7a. In this domain, two cubes, A and B, are defined as having four states: "A on top of B", "B on top of A", "A to the left of B", and "B to the left of A", as shown in Fig. 12.7b.

The task, in this case, is to create a transition of contact states between the two cubes. For instance, one transition is from the state A to the left of B to the state A on top of B. To each state transition we can assign one necessary operation that results in the desired next state. In this case, it is the operation Put-A-on-top-of-B. This association is defined as a *task model*.

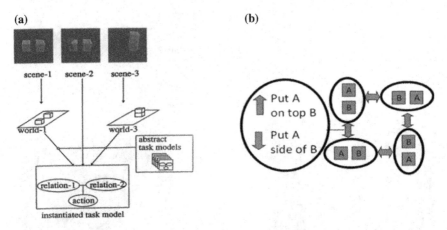

Fig. 12.7 Task recognition and state transitions in the two-block world. **a** Object recognition and task recognition. **b** States and state transitions in the two-block world

The purpose of task recognition is twofold. First, by dividing a continuous observation space into a discrete set of states and tasks, we can reduce the effect of observation errors. In this example, the object recognition result contained a small positional error. However, because the four states were adequately discrete, the result was correctly classified as one of the four states. A few error-correction examples in the polyhedral world can be found in Suehiro and Ikeuchi [9].

The second purpose is to separate observation from execution. The tasks obtained from observation are independent of the robot hardware; different robots can share the same observation module, and only task-mapping modules need to be specific to the robot hardware. We can make robots with different hardware execute the same set of tasks by simply replacing the mapping module without changing the observation module.

Under this task-skill paradigm, we employ the divide-and-conquer strategy to find the appropriate task domains that have the necessary and sufficient task sets. These domains include two cubes [7], two polyhedral objects [8], mechanical parts [10], and a knotting rope world [11].

12.2.2   Task Models for Manipulating Polyhedral Objects with Translation Motions

This section explains task recognition for handling polyhedral objects with translation motions. First, we pay attention to admissible moving directions of one polyhedral object manipulated, and we define a state by using its admissible movement directions. The movement of one polyhedral object, ΔT, is constrained by other environmental polyhedral objects. These constraints are represented by simultane-

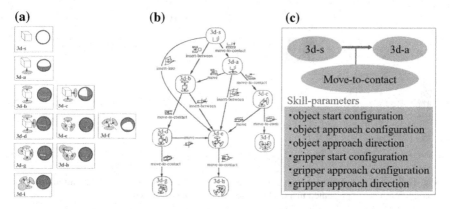

Fig. 12.8 Task model. **a** 10 face-contact states. **b** 13 admissible transitions and corresponding actions. **c** Abstract task model (example)

ous linear inequality equations.

$$\mathbf{N}_1 \cdot \Delta \mathbf{T} \geq 0$$
$$\mathbf{N}_2 \cdot \Delta \mathbf{T} \geq 0$$
$$\vdots$$
$$\mathbf{N}_n \cdot \Delta \mathbf{T} \geq 0,$$

where $\mathbf{N}_1, \mathbf{N}_2, \ldots, \mathbf{N}_n$ are surface normals at the contact points of those environmental polyhedral objects. Solutions of simultaneous linear inequality equations correspond to a set of movable directions. Solutions of all possible simultaneous linear inequality equations are characterized into 10 patterns [12]. Thus, for the polyhedral object world with translations, we can define 10 states as necessary and sufficient. Figure 12.8a depicts those 10 states defined by the solution areas.

Off-line, we set up abstract task models based on the possible transitions among these 10 states. For this, we enumerated possible transitions among them. Theoretically, if we have 10 states as starting states and 10 as ending states, we will have 100 possible transitions among them. By carefully enumerating these possibilities, however, we arrive at 13 admissible transitions. To each, we assign one robot action necessary to generate the transition, as shown in Fig. 12.8b. This pair of state transition and necessary action forms one abstract task model for manipulation.

To each abstract task model, we also prepared slots to characterize skill parameters necessary to complete the task such as grasping point, approach point, and prototypical trajectory. Figure 12.8c shows one example of abstract task models. This example depicts the move-to-contact task, which generates a transition from the *3d-s* state to the *3d-a* state. This task model has 6 slots for skill parameters. These skill parameters are collected, after an image segment is recognized as one particular task. Then, the corresponding task model is instantiated online.

(a)   **(b)**   **(c)**

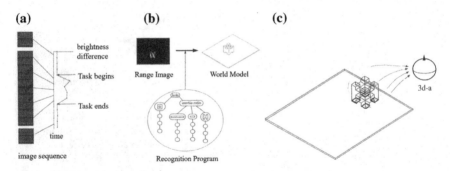

Fig. 12.9 Task recognition. **a** Image sequence to detect the end of a task demonstration. **b** Object recognition result represented in a CAD model. **c** Face-contact detection

Fig. 12.10 Robot execution of a peg-insertion task learned from human demonstrations

Online, the vision system continuously observes the scene. When a demonstrator conducts one action, a brightness disturbance occurs in the image sequence. From the onset and offset of the disturbance, the system detects the occurrence of one action and its duration. See Fig. 12.9a. At the offset of the disturbance, the system obtains the range image of the scene by using a light-strip range sensor: a class of active-lighting sensors. By subtracting the before-range image from the after-range image, the system determines the region corresponding to the object manipulated over the duration, recognizes which object is manipulated from the region, and represents the recognition results in a 3D simulation model to copy the current world(*object-recognition*) as shown in Fig. 12.9b.

In the simulation model, the system checks the face contact between this newly manipulated object against the previously existing environmental objects. See Fig. 12.9c. From the normal distribution of the contact faces, the system identifies the current state, determines the state transition by comparing it with the previous one, and identifies which task model to be instantiated. When one task model is instantiated, it collects necessary skill parameters from the stored image sequence, such as object location, grasping point, and approach direction. The system performs the corresponding robot actions by using the skill parameters collected (i.e., *task recognition*).

We can extend this method to include not only translation, but also rotation operations [11]. Figure 12.10 shows a performance sequence conducted by a humanoid robot. It shows the sequence of a robot execution of successfully performed peg-insertion operation learned from human demonstrations.

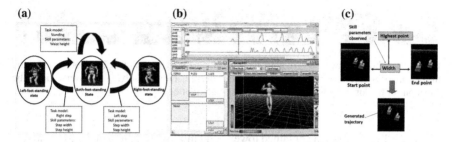

Fig. 12.11 Lower-body task model and skill parameters. **a** Task models and their transitions. **b** Task recognition using motion capture data. **c** Skill parameters

12.2.3 Lower-Body Task Models

We applied the task-skill paradigm to a humanoid robot for performing a Japanese folk dance [13]. This section focuses on modeling lower-body tasks for these dances.

Tasks for the lower body were defined with respect to contact states between the feet and the floor. The lower body has three contact states: left-foot contact, right-foot contact, and both-feet contact. Among these states, we defined three task transitions. Thus, three task models were created: right-step task, left-step task, and standing task. See Fig. 12.11a. For each task model, skill parameters were defined in a similar way to those of the polyhedral world.

In the online mode, human performance was recorded using a motion-capture system based on the active-lighting method. A dancer mounted 16 markers on his/her body, illuminated by infrared lights. They were observed by eight infrared cameras and provided a sequence of eight marker positions based on the triangulation method. The sequence was segmented into several segments, each of which corresponded to one task as shown in Fig. 12.11b.

After a segment was identified as one that performs a particular task, task recognition took place, and the corresponding task model was instantiated. Then, skill parameters of the task model were collected from the image segment. For example, if the segment was recognized as a right-step task from the distance between the floor and the left foot, the right-step task model was instantiated. Then, from the segment, by tracking the trajectory of the right foot, the skill parameters for this task model, step width and step height, were determined.

For each step task, a prototypical trajectory was also stored as a skill parameter in the task model. In this example, the prototypical trajectory of the right foot was retrieved from the slot in the task model. By using the skill parameters determined in the previous step (i.e., step width and step height), the system modified its prototypical trajectory to the one that satisfied these parameters. See Fig. 12.11c. After the foot trajectory was determined, by solving the inverse kinematic equations of the entire lower body, the system obtained the lower-body motions of the humanoid robot.

Fig. 12.12 Obtained lower body motion. First, we obtain the trajectory of the left foot by modifying the prototypical trajectory by the skill parameters observed. Then, by solving the inverse kinematic equations of the lower body, we obtain the lower-body motions

Figure 12.12 shows the obtained lower-body motion from human demonstrations. First, we obtained the trajectory of the left foot. Then, by solving the inverse kinematic equations of the lower body, we obtained the lower-body motions.

12.2.4   Upper-Body Task Models

We designed task models for upper-body motions based on Labanotation, a representation method used by the dance community to describe dances [14]. A dance or a gesture consists of a series of human motions. Such a continuous sequence can be described using Labanotation, which was developed by Rudolf V. Laban in the early twentieth century as a method of movement notation [15].

A Labanotation score is drawn in two dimensions (body columns and time rows) as shown in Fig. 12.13a. The vertical solid and dotted lines represent each body column. Each column, corresponding to one body part, contains Labanotation symbols, such as the rectangular and triangular symbols displayed in Fig. 12.13a. They represent the way each body part moves along the flow of time, which flows from bottom to top. Labanotation symbols are scaled to fit the starting and ending times, and the gap between two symbols in a column indicates a lack of motion in that period or holding of the previous pose during that period. The columns are divided into left and right sides, corresponding to the left and right sides of the body. These columns correspond to body parts. Figure 12.13a explicitly represents the upper arms, lower arms, and head. Other body parts, such as left and right feet and their support information, are omitted.

Each task model corresponds to one state transition of human performance, (i.e., spatial information of a Labanotation symbol). Each symbolic shape and color of Labanotation corresponds to one task model. Regarding skill parameters, in our current implementation, we only use the duration of one transition. Regarding trajectory and velocity of execution, we use a linear interpolation.

Figure 12.14 shows an overview of an online system. In our observation module, an active-lighting range sensor (i.e., Kinect) was used to record human motions. From skeleton data obtained using the depth sensor, key frame extraction and Labanotation encoding (task recognition) were conducted.

In the task-mapping module, Labanotation scores are converted into robot motions. A task-mapping module is specific to a robot configuration. A task only provides the start and end poses as represented by Labanotation symbols. For a robot motion, we

(a)

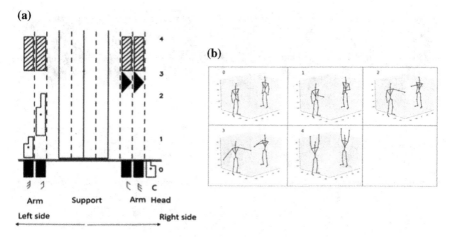

(b)

Fig. 12.13 Labanotation and corresponding poses. **a** Labanotation score representing upper-body parts. **b** Corresponding poses depicted using stick figures

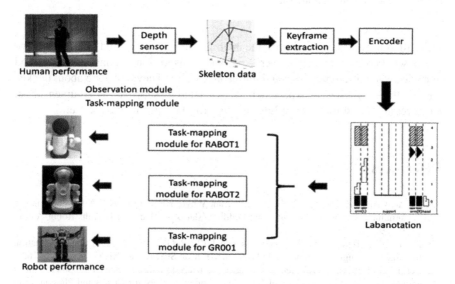

Fig. 12.14 Task recognition and mapping

require, as skill parameters, a trajectory to specify the intermediate motions between the two poses. We can thus implement a linear interpolation method for determining joint angles.

In doing so, we compared human motions with robot motions between two key frames, as shown in Fig. 12.15. The top and bottom rows show the original human and robot motions, respectively. The poses surrounded by the dotted boxes are the poses in the key frames. Because the motion speeds of the three robots differed, owing to

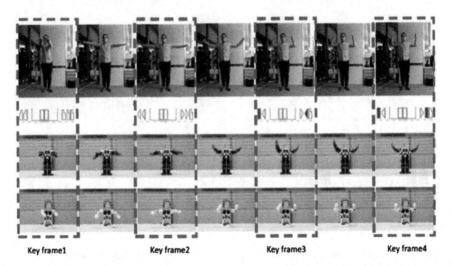

Fig. 12.15 Human and robot motions

their different control mechanisms, the intermediate poses captured between the two key poses were different. However, from visual inspection, human observers tend to neglect such differences during intermediate poses. They only care about the key poses in the key frames, as expected. In fact, this finding supports our argument about the necessity and sufficiency of Labanotation for the equivalent dance set.

References

1. Dissanayake MG, Newman P, Clark S, Durrant-Whyte HF, Csorba M (2001) A solution to the simultaneous localization and map building (slam) problem. IEEE Trans Robot Autom 17(3):229–241
2. Xue L, Ono S, Banno A, Oishi T, Sato Y, Ikeuchi K (2012) Global 3d modeling and its evaluation for large-scale highway tunnel using laser range sensor. SEISAN KENKYU 64(2):155–160
3. Besl PJ, McKay ND (1992) Method for registration of 3-d shapes. In: Sensor fusion IV: control paradigms and data structures, vol 1611. International Society for Optics and Photonics, pp 586–606
4. Oishi T, Kurazume R, Nakazawa A, Ikeuchi K (2005) Fast simultaneous alignment of multiple range images using index images. In: Fifth international conference on 3-d digital imaging and modeling (3DIM'05). IEEE, pp 476–483
5. Ono S, Ikeuchi K (2004) Self-position estimation for virtual 3d city model construction with the use of horizontal line laser scanning. Int J ITS Res 2:67–75
6. Bolles RC, Baker HH, Marimont DH (1987) Epipolar-plane image analysis: an approach to determining structure from motion. Int J Comput Vis (IJCV) 1(1):7–55
7. Ikeuchi K, Suehiro T, Tanguy P, Wheeler M (1991) Assembly plan from observation. In: Annual research review. The Robotics Institute, Carnegie Mellon University, pp 37–53
8. Ikeuchi K, Suehiro T (1994) Toward an assembly plan from observation. I. Task recognition with polyhedral objects. IEEE Trans Robot Autom 10(3):368–385

9. Suehiro T, Ikeuchi K (1992) Towards an assembly plan from observation: part II: correction of motion parameters based on fact contact constraints. In: Proceedings of the IEEE/RSJ international conference on intelligent robots and systems (IROS), vol 3. IEEE, pp 2095–2102

10. Ikeuchi K, Kawade M, Suehiro T (1993) Assembly task recognition with planar, curved and mechanical contacts. In: Proceedings of the IEEE international conference on robotics and automation (ICRA). IEEE, pp 688–694

11. Takamatsu J, Morita T, Ogawara K, Kimura H, Ikeuchi K (2006) Representation for knot-tying tasks. IEEE Trans Robot 22(1):65–78

12. Kuhn HW, Tucker AW (1956) Linear inequalities and related systems. In: Annals of mathematics studies. Princeton University Press

13. Nakaoka S, Nakazawa A, Kanehiro F, Kaneko K, Morisawa M, Hirukawa H, Ikeuchi K (2007) Learning from observation paradigm: leg task models for enabling a biped humanoid robot to imitate human dances. Int J Robot Res 26(8):829–844

14. Ikeuchi K, Ma Z, Yan Z, Kudoh S, Nakamura M (2018) Describing upper-body motions based on labanotation for learning-from-observation robots. Int J Comput Vis (IJCV) 126(12):1415–1429

15. Guest AH (2013) Labanotation: the system of analyzing and recording movement. Routledge

Index

A

Aberration, 39
Abstract task model, 297
Active BA, 256
Active stereo, 125
Amplitude-modulated laser, 265
Archeological analysis, 264
Area light sources, 96
Auto-calibration, 49, 249, 256

B

Back projection, 35
Balloon sensor, 273
Bayer pattern, 65
Bayon temple, 272
Bell-shaped, 95
Bi-directional reflectance distribution function (BRDF), 15, 183, 186, 214
Bi-directional texture function (BTF), 86
Bioluminescence, 93
Blackbody radiation, 90
Bolometers, 82
Brightness, 12
Brown–Conrady model, 39
Bundle-adjustment (BA), 251

C

Calibration objects, 39
Cambodian project, 263
Camera calibration, 39
Camera coordinate system, 32
Camera model, 31
Cardiac beat, 256

Charge-coupled-device (CCD), 63, 257
Circular polarization, 165, 166
Climbing sensor, 273
Coded aperture (CA), 175
Coherent light, 101
Color filters, 65
Complementary metal oxide semiconductor (CMOS), 63
Computer vision, 69
Coordinate transformation, 33
Coplanarity, 172
Correspondence problem, 126
Corresponding-point problem, 52
Cultural heritage, 263
Cyber archeology, 264
Cyrax sensor, 273

D

Dark current noise, 64
Data association, 291
de Bruijn sequence, 142, 144
Decorative tumulus, 264
Deep photometric stereo, 122
Degree of polarization, 79, 80, 163, 164, 166, 187
Depth camera, 70
Depth from defocus (DfD), 175–177
Diffractive optical elements (DOE), 101, 103, 243
Diffuse reflection, 7
Digital camera, 32
Digital micro-mirror devices (DMD), 99

© Springer Nature Switzerland AG 2020
K. Ikeuchi et al., *Active Lighting and Its Application for Computer Vision*,
Advances in Computer Vision and Pattern Recognition,
https://doi.org/10.1007/978-3-030-56577-0

Printed in the United States
by Baker & Taylor Publisher Services